Progress in Inflammation Research

Series Editor

Prof. Dr. Michael J. Parnham
PLIVA
Research Institute
Prilaz baruna Filipovica 25
10000 Zagreb
Croatia

Published titles:
T Cells in Arthritis, P. Miossec, W. van den Berg, G. Firestein (Editors), 1998
Chemokines and Skin, E. Kownatzki, J. Norgauer (Editors), 1998
Medicinal Fatty Acids, J. Kremer (Editor), 1998
Inducible Enzymes in the Inflammatory Response, D.A. Willoughby, A. Tomlinson (Editors), 1999
Cytokines in Severe Sepsis and Septic Shock, H. Redl, G. Schlag (Editors), 1999
Fatty Acids and Inflammatory Skin Diseases, J.-M. Schröder (Editor), 1999
Immunomodulatory Agents from Plants, H. Wagner (Editor), 1999
Cytokines and Pain, L. Watkins, S. Maier (Editors), 1999
In Vivo *Models of Inflammation*, D. Morgan, L. Marshall (Editors), 1999
Pain and Neurogenic Inflammation, S.D. Brain, P. Moore (Editors), 1999
Anti-Inflammatory Drugs in Asthma, A.P. Sampson, M.K. Church (Editors), 1999
Novel Inhibitors of Leukotrienes, G. Folco, B. Samuelsson, R.C. Murphy (Editors), 1999
Vascular Adhesion Molecules and Inflammation, J.D. Pearson (Editor), 1999
Metalloproteinases as Targets for Anti-Inflammatory Drugs, K.M.K. Bottomley, D. Bradshaw, J.S. Nixon (Editors), 1999
Free Radicals and Inflammation, P.G. Winyard, D.R. Blake, C.H. Evans (Editors), 1999
Gene Therapy in Inflammatory Diseases, C.H. Evans, P. Robbins (Editors), 2000
New Cytokines as Potential Drugs, S. K. Narula, R. Coffmann (Editors), 2000
High Throughput Screening for Novel Anti-inflammatories, M. Kahn (Editor), 2000

Forthcoming titles:
Inflammatory Processes. Molecular Mechanisms and Therapeutic Opportunities, L.G. Letts, D.W. Morgan (Editors), 2000
Novel Cytokine Inhibitors, G. Higgs, B. Henderson (Editors), 2000
Cellular Mechanisms in Airway Inflammation, C. Page, K. Banner, D. Spina (Editors), 2000

Immunology and Drug Therapy of Allergic Skin Diseases

Carla A. F. M. Bruijnzeel-Koomen
Edward F. Knol

Editors

Springer Basel AG

Editors

Prof. Dr. C.A.F.M. Bruijnzeel-Koomen
Department of Dermatology
University Hospital Utrecht
Heidelberglaan 100
NL-3584 CX Utrecht
The Netherlands

Dr. E. Knol
Department of Dermatology
University Hospital Utrecht
Heidelberglaan 100
NL-3584 CX Utrecht
The Netherlands

A CIP catalogue record for this book is available from the Library of Congress, Washington D.C., USA

Deutsche Bibliothek Cataloging-in-Publication Data
Immunology and drug therapy of allergic skin diseases /
ed. by C.A.F.M. Bruijnzeel-Koomen, E.F. Knol - Basel ; Boston ; Berlin : Birkhäuser, 2000
    (Progress in inflammation research)

ISBN 978-3-0348-9579-8   ISBN 978-3-0348-8464-8 (eBook)
DOI 10.1007/978-3-0348-8464-8

Originally published by Birkhäuser Verlag, Basel, Switzerland in 2000

Printed on acid-free paper produced from chlorine-free pulp. TCF ∞
Cover design: Markus Etterich, Basel
Cover illustration: Edward F. Knol, Utrecht. The figure is a haematoxylin-eosin staining of a formalin-fixed paraffin-embedded biopsy of human skin after intracutaneous injection of the chemokine RANTES demonstrating leukocyte infiltration in the dermis.

9 8 7 6 5 4 3 2 1

Contents

List of contributors

Paul J. Baselmans, Research Institute Jouveinal, Parke-Davis, 94265 Fresnes, France

Jan D. Bos, Department of Dermatology A0-235, Academic Medical Center, University of Amsterdam, P.O. Box 22700, 1100 DE Amsterdam, The Netherlands; e-mail: j.d.bos@amc.uva.nl

Carla A.F.M. Bruijnzeel-Koomen, Department of Dermatology/Allergology, G02-124, University Medical Center Utrecht, P.O. Box 85.500, 3508 GA Utrecht, The Netherlands; e-mail: m.huisman@digd.azu.nl

Anton C. de Groot, Department of Dermatology, Carolus-Liduina Ziekenhuis, P.O. Box 1101, 5200 BD 's-Hertogenbosch, The Netherlands; e-mail: anton.de.groot-huidarts@wxs.nl

Gerald R. Dubois, Research Institute Jouveinal, Parke-Davis, 94265 Fresnes, France; e-mail: gerald.dubois@wl.com

Jörn Elsner, Hannover Medical University, Department of Dermatology and Allergology, Ricklinger Str. 5, 30449 Hannover, Germany; e-mail: jelsner@compuserve.com

Peter S. Friedmann, Dermatopharmacology Unit, Level F, South Block, Southampton General Hospital, Southampton, SO16 6YD, UK; e-mail: psf@soton.ac.uk

Clive E. Grattan, Dermatology Centre, West Norwich Hospital, Norwich NR2 3TU, UK; e-mail: clive.grattan@norfolk-norwich.nhs.com

Alexander Kapp, Hannover Medical University, Department of Dermatology and Allergology, Ricklinger Str. 5, 30449 Hannover, Germany

Edward F. Knol, Department of Dermatology/Allergology, G02-124, University Medical Center Utrecht, P.O. Box 85.500, 3508 GA Utrecht, The Netherlands; e-mail: e.f.knol@digd.azu.nl

Hans F. Merk, Department of Dermatology and Allergology, University Hospital, RWTH Aachen, Pauwelsstrasse 30, 52074 Aachen, Germany; e-mail: hans.merk@post.rwth-aachen.de

Geert C. Mudde, Research Institute Jouveinal, Parke-Davis, 94265 Fresnes, France

Birgit A. Pees, Dermatopharmacology Unit, Level F, South Block, Southampton General Hospital, Southampton, SO16 6YD, UK

Cornelis J.W. van Ginkel, Department of Dermatology/Allergology, University Hospital, P.O. Box 85500, 3508 GA Utrecht, The Netherlands; e-mail: cjginkel@xs4all.nl

Els Van Hoffen, Department of Dermatology/Allergology (G02.124), University Medical Center Utrecht, P.O. Box 85.500, 3508 GA Utrecht, The Netherlands; e-mail: e.vanhoffen@lab.azu.nl

Frank C. Van Reijsen, Utrecht, The Netherlands; e-mail: reijsen@xs4all.nl

Preface

Allergic skin diseases belong to the most frequent dermatoses. This book deals with both fundamental, in particular immunological aspects, as well as clinical symptoms and therapeutic strategies of allergic skin diseases.

Allergic skin diseases comprise a variety of players in its pathophysiological mechanism. Its distinct immunologic characteristics have initiated the designation "skin immune system". In this system the cells involved in the local immune response of the skin are described. The direct cell-to-cell interactions will be described in greater detail in a separate chapter. Moreover, individual chapters each deal with T lymphocytes, mast cells and eosinophils, cells involved in the pathomechanisms of allergic skin disease.

The allergic skin diseases which will be discussed are atopic dermatitis, being the chronic inflammatory skin disease with the highest prevalence, allergic contact dermatitis with special focus on contact dermatitis from cosmetics, being the most frequent cause of contact dermatitis and occupational contact dermatitis. Urticaria is not an allergic skin disease *per se*, but is an important symptom of an allergic/anaphylactic reaction. The broad etiology of urticaria and the value of several therapeutic strategies will be explained.

This book is meant for clinicians working in the fields of dermatology, allergology, pediatrics and even general practitioners who will often be the first to be confronted with allergic skin diseases. This overview of both the clinical and pathophysiological characteristics of allergic skin diseases will also be of benefit to researchers active in the field of allergic skin diseases.

Edward F. Knol
Carla A.F.M. Bruijnzeel-Koomen

Skin immune system (SIS)

Jan D. Bos

Department of Dermatology AO-235, Academic Medical Center, University of Amsterdam, P.O. Box 22700, 1100 DE Amsterdam, The Netherlands

Introduction

The immune system is complex and as a result, its dysregulations are even more difficult to understand. Atopy is a syndrome with apparent immune dysregulation. And so are atopic skin diseases, especially atopic dermatitis or eczema [1, 2]. Knowledge of normal immune function is necessary in order to understand aberrations such as atopy. This also is true for the skin where a good understanding of atopic skin diseases is only possible with adequate knowledge of the normal immune function of skin. In this chapter, the skin immune system (SIS) will be described. The SIS embodies the set of immune-response associated cellular and molecular elements of the human integument. Together they act as the physiological immune defense system of skin. There are numerous dysregulations of SIS. Those abnormalities in atopic skin diseases that can be directly related to the components of SIS will be referred to.

Systemic immunity and atopy

One may categorize the organs of the human body that are involved in the generation of immune responses into primary, secondary and tertiary immune organs. Primary immune organs in man are bone marrow and thymus, where the cellular elements of the immune system are formed. Secondary immune organs are the spleen and the lymph nodes, where naïve cells become primed and ready for secondary responses. Memory cells as well as antibodies are produced in these secondary immune organs, and they are dispersed over the tissues through the peripheral circulation.

Tertiary immune organs are the organs where effector immune responses occur. It is without doubt that skin is a tertiary immune organ. It has been speculated that skin might have a secondary or even primary immune function, especially early in life [3]. In the context of this chapter, this will not be further detailed. The fact that a given organ, such as skin, is a tertiary immune organ, does not exclude it from being the site of initiation of new immune responses. In these cases however, the actual memorization occurs in the secondary immune organs, the skin-draining lymphoid organs.

The genetic basis of atopy is now under active investigation, and the possible genes and gene polymorphisms identified thus far mainly point to abnormalities in cellular proteins associated with immune responses. For example, the gain of function mutation in the interleukin (IL)-4 receptor (R) α-chain is a polymorphism highly associated with atopy and with hyper-IgE syndrome [4]. It seems reasonable to assume that this abnormality is present as soon as the IL-4R α-chain is expressed on lymphocytes. Another polymorphism found to be associated with atopic dermatitis but not with other atopic diseases is that of the enzyme mast cell chymase [5]. Again, it seems reasonable to assume that this polymorphism becomes expressed as soon as mast cell precursors start to express the enzyme in their cytoplasm.

It is thus believed that the genetic basis of atopy becomes expressed with the differentiation of immature, precursor (stem) cells into more specialized cells, concurrent with mRNA transcription and formation of these differentiation-associated proteins.

The concept of the skin immune system (SIS)

As indicated above, understanding the normal function of skin is a prerequisite for understanding dysregulation. In that context, it is important to realize that skin is more than a physical defense barrier. Other well-known functions are its roles in the maintenance of body temperature, regulation of stable circulation, production of endocrine mediators and the bearing of peripheral neural receptors and nerve endings. The skin thus serves as the central nervous system's largest outpost and is a major sensory organ. Psychological and social functions are also evident.

It must be assumed that, during evolution, the defense role of the integument has become gradually more complex and differentiated. Human skin has thus obtained innate as well as acquired immunity-related cellular components as well as molecular defense systems (Tab. 1). If one dissects the skin into its cellular elements, and divides these over those cells that are directly immune-response related and those that are not (Tab. 2), it becomes clear that immunological mechanisms are of considerable importance for the general function of skin. This observation formed the basis for the gathering of the immune components of normal human skin into the designation "Skin immune system" [6], first published in 1986.

Others have also tried to give the immune function of skin its due by giving it a special name. Streilein, earlier in 1978, coined the term "skin-associated lymphoid tissues" (SALT) to include keratinocytes, intra-epidermal Langerhans cells (LCs) as antigen-presenting cells, skin-seeking T lymphocytes, endothelial cells of the skin

Table 1 - Cellular and molecular components of the skin immune system (SIS)

Cellular constituents	Humoral constituents
Keratinocytes	Antimicrobial peptides, defensins
Dendritic antigen presenting cells	Complement and complement regulatory proteins
Monocytes/macrophages	Immunoglobulins
Granulocytes	Cytokines, neuropeptides
Mast cells	Fibrinolysins
Lymphatic/vascular endothelial cells	Eicosanoids and prostaglandins
T lymphocytes	Free radicals

modified after [32]

directing these skin-seeking cells into the dermis, and the skin-draining lymph nodes, being the specific localization of induction of immunity by antigens that have been processed and transported by LCs [7].

Since the introduction of the SALT concept, some investigators have entirely focused on the epidermis and suggested it to be an immunological organ because of its combination of keratinocytes, dendritic cells, and T lymphocytes. Obviously, concepts focusing solely on the epidermis are incomplete as they exclude the major site of immunological action in skin. The preferential distribution of T cells, monocytes and most other cellular constituents of the skin immune system is in the dermis, especially in its papillary part.

Thus, Sontheimer (in 1989) gave his definition of the dermal microvascular unit (DMU), which was to point to the very center of immunological reactivity in most inflammatory and immunologically mediated dermatoses [8]. Directly around the postcapillary venules, we find accumulations of T cells, monocytes and tissue macrophages, mast cells, and dendritic cells. All elements of immune reactivity are present and it is no surprise that most inflammatory skin diseases show expansion of the cellular elements of the DMU. Thus, DMU might be considered to be a subsystem of SIS.

Nickoloff, in 1993, proposed the term dermal immune system (DIS) to be the cellular and humoral counterpart of SALT [9]. It included fibroblasts, mainly because they are intrinsically related to homeostasis of other skin components, such as epidermis. With the exception of SALT's lymph nodes and DIS's fibroblasts, the two concepts SALT and DIS might also be considered as functional subsystems of SIS.

From a quantitative point of view, keratinocytes and fibroblasts are the key structural cellular components of the integument. But Table 2 clearly shows that qualitatively, about half of the cellular constituents of normal human skin are cells

Table 2 - Overview of the cells of normal human skin, divided over immune-response associated and non-immune response-associated cells

Immune response-associated	Not immune-response associated
Keratinocytes	Merkel cells
Langerhans cells (epithelial dendritic cells)	Melanocytes
Tissue dendritic cells (dermal dendrocytes)	Fibroblasts/fibrocytes/myofibroblasts
Tissue macrophages and monocytes	Pericytes
Granulocytes	Eccrine gland and duct cells (acrosyringium)
Mast cells	Apocrine gland and duct cells
T lymphocytes and subpopulations	Sebocytes
Vascular endothelial cells	Schwann cells
Lymphatic endothelial cells	Smooth muscle cells

modified after [32]

involved in immune processes, emphasizing the complex role of skin in the expression of immune diseases.

Since the pathogenesis of atopic dermatitis is best described at the cellular level,
it seems of importance that the immune response-associated cellular components of
SIS are discussed in some detail. In other words, the immunopathogenesis of atopic
eczema and perhaps other atopic skin diseases seems best explained at the level of
cellular pathology. Thus, a description of the normal cellular constituents of SIS will
follow here.

Keratinocytes

Keratinocytes form the major constituent of the epidermis. Their main role is to differentiate into corneocytes, keratin-rich envelopes that form the corneal layer, the
primary and highly effective structural defense layer of the skin. But keratinocytes
have other biological functions. In wound healing, they show an impressive regen-
erative capacity, allowing them to efficiently close large de-epidermized areas if necessary. As to their role in immunity, they may function as accessory cells in antigen
and allo-antigen presentation [10].

Most importantly, they are able to synthesize a wide variety of cytokines. Non-
specific stimulation, such as skin irritation with detergents or exposure to ultraviolet irradiation, leads to the production by keratinocytes of proinflammatory
cytokines such as IL-1 and IL-6. This pro-inflammatory response is thought to give
the epidermis a specialized function in the preparation of the skin to subsequent,

more specific defense reactions [11]. These newly synthesized cytokines induce increased expression of adhesion molecules on vascular endothelial cells, resulting in an accelerated influx of leukocytes from the peripheral blood into the skin tissues. Among these leukocytes, there are antigen-specific T cells that are thus recruited into a damaged skin area.

It is at the level of cytokine production by keratinocytes that atopic dermatitis patients have shown intriguing abnormalities. Atopic patients have keratinocytes that express high amounts of granulocyte-macrophage colony-stimulating factor (GM-CSF), both at protein and mRNA level [12]. Keratinocytes cultured from atopic dermatitis patients spontaneously produce increased amounts of GM-CSF, IL-1α, and perhaps also IL-1ra (receptor antagonist) and tumor necrosis factor α (TNFα) [13].

These findings from Girolomoni's group have two distinct implications. First, by virtue of their constitutive increased production of pro-inflammatory cytokines, they might prepare atopic skin continuously for increased inflammatory and immune reactions. Thus, this might form the biological basis for the "irritability" of skin in atopic dermatitis patients. Non-specific activation of keratinocytes leads to even more enhanced pro-inflammatory reactions, leading to non-specific inflammation and exacerbation of atopic eczema itself. A second implication of these findings is that keratinocyte-derived GM-CSF together with Th2-derived IL-4 might be responsible for increased maturation of dendritic cells in the skin (dendritic cell lines can be generated from precursor cells using these two cytokines), and thus for increased antigen-presenting capacity in atopic skin lesions.

Langerhans cells (LCs) and other dendritic antigen presenting cells (APCs)

Langerhans cells (LCs) are seen as immature dendritic APCs with a tendency to home into epithelial tissues, most notably the epidermis. They are thought to be derived from precursors introduced into peripheral blood from the bone marrow. In the dermis, they have the phenotype of tissue dendritic cells (dermal dendrocytes). Epidermal LCs are believed to continuously trap antigens from the epidermal milieu, and take these through the lymph vessels to the T cell area's of the draining lymph nodes. During this passage, they are called veiled cells and are believed to mature into dendritic APCs, which by virtue of their special constellation of membrane-bound and secretory molecules, are able to prime naïve T cells in the lymphoid tissue (secondary immune organs).

The phenotype of epidermal LCs, but more apparently of cultured LCs, fits with strong APC function, and cultured LCs are thought to best represent matured DCs as they arrive in the lymph nodes [14]. It is believed that the dendritic cells of SIS are involved in antigen presentation in atopic skin diseases. The discovery of IgE molecules on LCs in atopic dermatitis skin by Bruijnzeel-Koomen [15] in 1986, has since led to the knowledge that LCs express high affinity receptors for IgE (FcεRI) and under certain circumstances also the low affinity receptor for IgE (FcεRII). The presence of (allergen-specific) IgE on these APC in skin is thought to be associated with a phenomenon of antigen focusing or facilitation of antigen presentation. It takes only minute amounts of allergens to be effectively presented to the T cells in the skin when such an IgE-mediated focusing on APCs has occurred [16].

Monocytes and macrophages

Normal human skin contains substantial amounts of monocytes and their more differentiated tissue counterparts, macrophages. There is a directly subepidermal accumulation of these cells, as well as a preferential distribution around the postcapillary venules of the superficial and peri-adnexal vascular plexus. In addition, they are sparsely but widely scattered in the dermis. They do not occur in the epidermis, with the exception of UVB-exposed human skin, where epidermal macrophages (CD1a$^-$, CD11b$^+$) appear 2 to 3 days following irradiation (Cooper cells or UV-macrophages) [17].

Tissue macrophages may be seen as the debris containers of the skin. When they contain huge amounts of erythrocyte debris (such as in vasculitis), they are called siderophages, and when they contain high amounts of melanosomes (such as in any melanocyte damaging epidermo-dermatitis), melanophages. In addition to their physiological role in cleaning up the cutaneous sites from degraded cells and tissue components during the constant renewal of skin and after microscopic injury, they are able to act as antigen-presenting cells. However, unlike DCs, they are not effective enough in inducing primary responses in naive cells, but they can stimulate memory T cells.

The exact interrelations between monocytes, macrophages, tissue dendritic cells (dermal dendrocytes), veiled cells and Langerhans cells in skin are difficult to summarize. A general distinction between monocytes/macrophages on the one hand and dendritic cell subpopulations on the other seems the most practical approach [18].

In atopy, an abnormality of the enzyme cyclic adenosine monophosphate-phosphodiesterase has been described. Its activity was found to be increased in monocytes derived from peripheral blood of atopy patients. Such increased activity may be associated with changed patterns of leukotriene (LT) and prostaglandin (PG) production [19].

Mast cells

Mast cells form normal constituents of the SIS. They are preferentially localized around the postcapillary venules of the superficial vascular plexus high in the papillary dermis. But they are also widely distributed throughout the connective tissue of the dermis. A distinction has been made in rats between connective tissue mast cells and mucosal mast cells. However, such a distinction has not been reproduced in human tissues. But mast cells may show heterogeneity, both morphologically, histochemically and functionally [20].

Mast cells are complex cells that respond to a wide variety of activation signals with a wide variety of mediators, that are either preformed and immediately released or synthesized after proper stimulation. And these mediators then again have a wide variety of biological effects. In atopy, an important signaling pathway is obviously through coupling of two FcεR1-bound allergen-specific IgE molecules upon binding of allergenic peptide. This forms the basis for intracutaneous challenge with allergens in individuals suspected of being atopic. It also is the basis for understanding urticarial reactions, such as after food allergenic challenge, that are so common in atopic individuals. The precise role of the polymorphism in mast cell chymase, as described above, remains to be established.

Since mast cells form the subject of an independent chapter in this book, readers are referred to that chapter for further details of these important cells.

Granulocytes

Granulocytes are bone-marrow derived specialized leukocytes of which three types exist. Neutrophilic granulocytes are short-lived cells of which billions are produced daily. They circulate in the peripheral blood but are not present in normal human skin, except of course intravascularly. Neutrophils are a very common cell type in many inflammatory conditions of the skin. They have a role as phagocytic cells and are considered to be part of the innate system of immune responses. In atopic dermatitis, they are seen when secondary infections have occurred. Basophilic granulocytes are related to mast cells. They may be observed in low numbers in involved atopic skin. They are not normally present in human skin. Eosinophilic granulocytes are allergy-associated and are dependent on the Th2 cytokine IL-5. Since eosinophilic granulocytes form the subject of an independent chapter in this book, readers are referred to that chapter for further details of these important cells.

Endothelial cells

The lymphatic and more importantly the vascular endothelial cells have important functions in skin immune reactions. They have functions in wound healing, coagulation, inflammation and immune reactions. Especially the trafficking of leukocytes in and out of the dermis is regulated by interaction of different molecules on leukocytes and their ligands (counterparts) on endothelial cells. Different stages are discerned in the process of leukocyte immigration. In the rolling stage, rapid de-accel-eration of passing leukocytes occurs primarily through the selectin family of adhe-sion molecules (P- and E-selectins). In later stages, interaction with integrins, inter-cellular adhesion molecule-1 (ICAM-1), and vascular cell adhesion molecule-1 (VCAM-1) leads to arrest of the cells. Diapedesis subsequently depends on other molecules such as CD31.

Not much is known about the possible abnormalities at endothelial cell level in atopy. One might assume that endothelial cells function normally in atopic skin dis-eases, only responding as they always do to the many mediators excreted by ker-atinocytes, mast cells, eosinophilic granulocytes and T cells in the atopic inflamma-tory process.

T cells

Normal human skin contains substantial numbers of T cells. In fact, there are more of them in the skin than in the peripheral blood of a normal individual [21]. They are mainly localized around the postcapillary venules of the superficial vascular plexus high in the papillary dermis, as well as around the postcapillary venules of the peri-adnexal vasculature. In dermatopathological reports, they are, together with the monocytes/macrophages and mast cells that also preferentially localize at these sites, often described as small perivascular infiltrates, which they are, but in small numbers they are physiological.

Dermal T cells express the classic markers CD2 and CD5. They are equally divided over CD4 and CD8 subpopulations. Most of them seem to be memory cells as they are HLA-DR+, CD25+, CD45RO+ and CD7-. About a third express the skin homing marker cutaneous lymphocyte antigen (CLA) [22]. In addition, T cells may be encountered in the normal epidermis, be it in very small numbers. They account for about 2% of the total number of T cells in normal human skin. They are CD2+, CD5+, with a preponderance for CD8 over CD4. The majority show a memory phe-notype with expression of CD45RO, but is insecure whether they show signs of recent activation.

A human counterpart for the dendritic epidermal T cells (DETCs), such as they occur in mice, has not been found [23]. DETCs express the $\gamma\delta$-type of T-cell recep-tor (TCR). In human skin, TCR-$\gamma\delta$+ cells form between 2 and 15% of the total pop-ulation, the remaining T cells expressing the more common TCR-$\alpha\beta$. But in contrast to the situation in mouse epidermis, human epidermal TCR-$\gamma\delta$+ cells are not den-dritic and they are rare. In mice, they form the majority of epidermal T cells.

In addition to the immunophenotypic distinction of T cells into various subpop-ulations, T cells may be subdivided functionally into subpopulations that produce different sets of cytokines [24, 25]. This concept of T-helper cell polarization from intermediate Th0 cells into either Th1 or Th2 cells emerged early in the 1990s, and has found substantial importance in understanding the pathogenesis of a number of important disorders, including atopy. In fact, it seems that the first human Th2 cells were identified in patients suffering from atopic dermatitis [26, 27].

The polarization towards Th2 cells in atopy forms the explanation for the tendency of atopic patients to produce allergen-specific IgE. The effects of Th2 cytokines IL-4 and IL-13 on B cells, together with cell membrane-bound interaction between CD40 and its ligand on B cells, leads to switching of μ- (IgM) to ε- (IgE) chain production. This contrasts with the normal situation, where the combined cytokines from Th0 and Th2 cells lead to switching of μ- (IgM) to γ (IgG).

In normal human skin, it is as yet undecided to what functional subpopulations the T cells belong. In atopy, Th2 type of cells are clearly involved in IgE production in the secondary immune (lymphoid) organs. In atopic dermatitis, Th2 are thought to be important in the early development of lesions [28]. But naturally, they do not play a role in the switch of μ- to ε-chain production in B cells, since B cells nor plasma cells are present in normal or atopic dermatitis skin [29]. The precise role of Th1, Th2 and Th0 cells within atopic dermatitis involved skin still remains to be established.

Finally, it is of interest to note that in one study, it was found that patients with "intrinsic" non-atopic dermatitis have T cells spontaneously producing high amounts of IL-5 but not IL-4, while T cells from atopic dermatitis patients produced both cytokines [30]. Thus, this might be the beginning of an immunological explanation for the existence of atopic dermatitis-like disease in non-atopics, i.e. in individuals in whom allergen-specific IgE cannot be detected.

Atopic dermatitis is usually clinically defined. A substantial number (10–20%) of patients with non-atopic atopic dermatitis-like dermatitis are included in studies when these clinical criteria are used. We have recently suggested to define the disease, at least in part, on the basis of its immunological characteristic, i.e. the presence of allergen-specific IgE, which is in itself T cell directed [31]. In this way, one can exclude these non-atopic atopic dermatitis-like dermatitis patients from future studies into the genetics, pathogenesis, clinical epidemiology and therapy of this common and devastating disease, and that is important because these non-atopic patients probably confound the outcome of such investigations.

Summary and conclusions

The human skin harbors a wide variety of immune-response associated cellular and humoral components, that together form the Skin immune system (SIS). It is the biological counterpart of the skin's physical defense function. Under normal conditions, the elements of SIS work together in physiological processes such as immunosurveillance and the invisible elimination of potentially sensitizing remains of degraded tissue components. SIS is able to do this without clinically visible signs or symptoms,

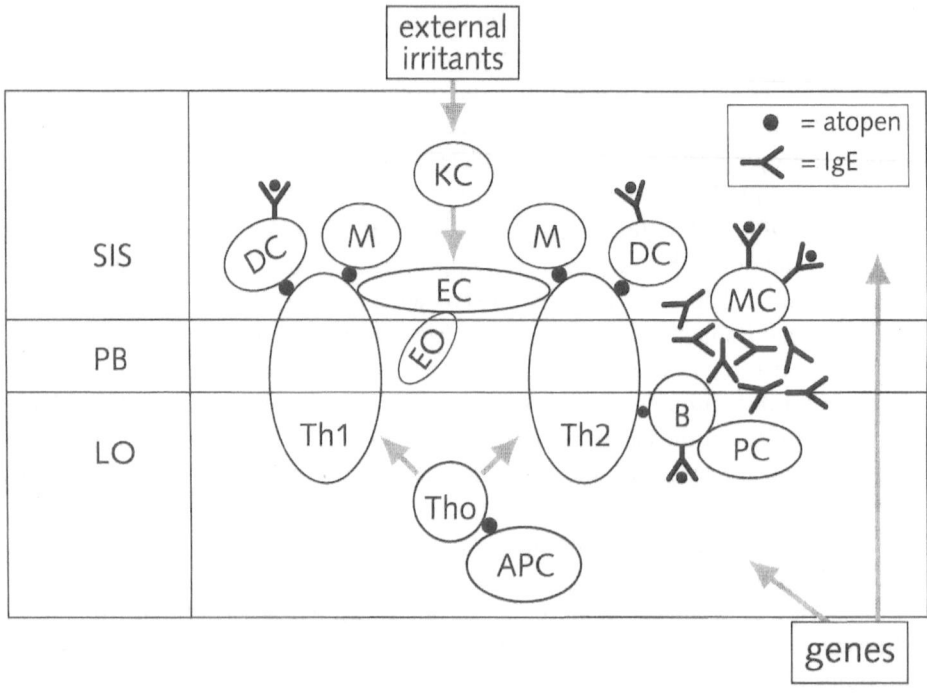

Figure 1
Overview of SIS and cellular pathology of atopic dermatitis
In the lymphoid organs (LO), the genetically determined preferential outgrowth of Th2 type lymphocytes occurs under the influence of antigen presenting cells (APC), leading to stimulation of B cells (B) and their differentiation into IgE secreting plasma cells (PC). IgE leaks into the peripheral blood (PB) and sensitizes the skin immune system (SIS) by binding to receptors on mast cells (MC) and dendritic cells (DC). Sensitization of MC seems essential for atopy-related urticaria and itch. Sensitization of DC leads to antigen focusing and extremely efficient antigen presentation within the skin. Nonspecific keratinocyte (KC) stimulation leads to a pro-inflammatory response, allowing upregulation of adhesion molecules on endothelial cells (EC) and influx of T cells as well as eosinophilic granulocytes (EO) into the skin. A complex interplay of SIS cells then occurs, ultimately leading to what is clinically seen as atopic dermatitis. (Modified from [2].)

although some itch or erythema is noted by every normal person on a daily basis. It is evident that a complexity such as SIS may become dysregulated by a wide variety of abnormalities. Such is the case in atopic skin diseases, where a variety of immune abnormalities are known to exist, each affecting the normal function of SIS in a different way, but together leading to clinically manifest dermatological disease.

The major cells thought to play a role in the pathogenesis of atopic dermatitis and other atopic skin diseases are the mast cells, eosinophils, dendritic cells and T cells.

It seems that a number of genetically determined primary and secondary immune organ abnormalities form the basis of the atopic syndrome. Some of these abnormalities subsequently affect the tertiary immune organ skin. As a result, the skin becomes a target of abnormal immune responses, clinically manifest as different forms of atopy-related skin diseases. In Figure 1 (a variant of its originally published version [2]) is shown the complex interrelationship of the cellular components affected by atopy, with their respective localizations in the lymphoid organs, the peripheral blood, and in the SIS.

References

1   Bos JD, Wierenga EA, Sillevis Smitt JH, Van der Heijden FL, Kapsenberg ML (1992) Immune dysregulation in atopic eczema. *Arch Dermatol* 128: 1509–1512

2   Bos JD, Kapsenberg ML, Sillevis Smitt JH (1994) Pathogenesis of atopic eczema. *Lancet* 343: 1338–1341

3   Fichtelius KE, Groth O, Liden S (1970) The skin, a first level lymphoid organ? *Int Arch Allergy* 37: 607–620

4   Hershey GKK, Friedrich MF, Esswein LA, Thoman ML, Chatila TA (1997) The association of atopy with a gain-of-function mutation in the α subunit of the interleukin-4 receptor. *N Engl J Med* 337: 1720–1725

5   Mao XQ, Shirakawa T, Yoshikawa T, Yoshikawa K, Kawai M, Sasaki S, Enomoto T, Hashimoto T, Furuyama J, Hopkin JM et al (1996) Association between genetic variants of mast-cell chymase and eczema. *Lancet* 348: 581–583

6   Bos JD, Kapsenberg ML (1986) The skin immune system (SIS): its cellular constituents and their interactions. *Immunol Today* 7: 235–240

7   Streilein JW (1978) Lymphocyte traffic, T cell malignancies and the skin. *J Invest Dermatol* 71: 167–171

8   Sontheimer RD (1989) Perivascular dendritic macrophages as immunobiological constituents of the human dermal perivascular unit. *J Invest Dermatol* 93: 96S–101S

9   Nickoloff BJ (ed) (1993) *Dermal immune system.* CRC Press, Boca Raton

10   Chu AC, Morris JF (1997) The Keratinocyte. In: JD Bos (ed): *Skin immune system (SIS): Cutaneous immunology and clinical immunodermatology.* CRC Press, Boca Raton, 43–57

11   Barker JNMN, Mitra RS, Griffiths CEM, Dixit VH, Nickoloff BJ (1991) Keratinocytes as initiators of inflammation. *Lancet* 337: 211–214

12   Pastore S, Fanales-Belasio E, Albanesi C, Chinni LM, Giannetti A, Girolomoni G (1997) Granulocyte macrophage colony-stimulating factor is overproduced by keratinocytes in

atopic dermatitis. Implications for sustained dendritic cell activation in the skin. *J Clin Invest* 99: 3009–3017

13   Pastore S, Corinti C, La Placa M, Didona B, Girolomoni G (1998) Interferon-γ promotes exaggerated cytokine production in keratinocytes cultured from patients with atopic dermatitis. *J Allergy Clin Immunol* 101: 538–544

14   Teunissen MBM, Kapsenberg ML, Bos JD (1997) Langerhans cells and related skin dendritic cells. In: JD Bos (ed): *Skin immune system (SIS): Cutaneous immunology and clinical immunodermatology.* CRC Press, Boca Raton, 59–83

15   Bruijnzeel-Koomen C, Van Wichen DF, Toonstra J, Berrens L, Bruijnzeel PLB (1986) The presence of IgE molecules on epidermal Langerhans cells in patients with atopic dermatitis. *Arch Dermatol Res* 278: 199–205

16   Van der Heijden FL, Van Neerven RJJ, Van Katwijk M, Bos JD, Kapsenberg ML (1993) Serum-IgE-facilitated allergen presentation in atopic disease. *J Immunol* 150: 3643–3650

17   Cooper KD, Oberhelman L, Hamilton TA, Baadsgaard O, Terhune M, LeVee G, Anderson T, Koren H (1992) UV exposure reduces immunization rates and promotes tolerance to episutaneous antigens in humans: relationship to dose, CD1a⁻ DR⁺ epidermal macrophage induction, and Langerhans cell depletion. *Proc Nat Acad Sci* 89: 8497–8501

18   Rowden G (1997) Macrophages and dendritic cells in the skin. In: JD Bos (ed): *Skin immune system (SIS): Cutaneous immunology and clinical immunodermatology.* CRC Press, Boca Raton, 109–146

19   Chan SC, Li S-H, Hanifin JM (1993) Increased interleukin-4 production by atopic mononuclear leukocytes correlates with increased cyclic adenosine monophosphate-phosphodiesterase activity and is reversible by phosphodiesterase inhibition. *J Invest Dermatol* 100: 681–684

20   Van Loveren H, Redegeld F, Matsuda H, Buckley Th, Teppema JS, Garssen J (1997) Mast cells. In: JD Bos (ed): *Skin immune system (SIS): Cutaneous immunology and clinical immunodermatology.* CRC Press, Boca Raton, 159–184

21   Bos JD, Zonneveld I, Das PK, Krieg SR, Van der Loos ChM, Kapsenberg ML (1987) The skin immune system (SIS): distribution and immunophenotype of lymphocyte subpopulations in normal human skin. *J Invest Dermatol* 88: 569–573

22   Foster CA, Elbe A (1997) Lymphocyte subpopulations of the skin. In: JD Bos (ed.): *Skin immune system (SIS): Cutaneous immunology and clinical immunodermatology.* CRC Press, Boca Raton, 85–108

23   Bos JD, Teunissen MBM, Cairo I, Krieg SR, Kapsenberg ML, Das PK, Borst J (1990) T cell receptor gamma/delta bearing cells in normal human skin. *J Invest Dermatol* 94: 37–42

24   Pène J, Rousset F, Brière F, Chrétien I, Bonnefoy JY, Spits H, Yokota T, Arai N, Banchereau J, De Vries JE (1988) IgE production by normal human lymphocytes is induced by interleukin 4 and suppressed by interferons γ and α and prostaglandin E2. *Proc Natl Acad Sci USA* 85: 6880–6884

25 Kapsenberg ML, Wierenga EA, Bos JD, Jansen HM (1991) Functional subsets of allergen-reactive human CD4$^+$ T-cells. *Immunol Today* 12: 392–395

26 Wierenga EA, Snoek M, De Groot C, Chretien I, Bos JD, Jansen HM, Kapsenberg ML (1990) Evidence for compartmentalization of functional subsets of CD4$^+$ T lymphocytes in atopic patients. *J Immunol* 144: 4651–4656

27 Van der Heijden FL, Wierenga EA, Bos JD, Kapsenberg ML (1991) High frequency of IL-4-producing CD4+ allergen-specific T lymphocytes in atopic dermatitis lesional skin. *J Invest Dermatol* 97: 389–394

28 Thepen T, Langeveld-Wildschut EG, Bihari IC, van Wichen DF, van Reijsen FC, Mudde GC, Bruijnzeel-Koomen CA (1996) Biphasic response against aeroallergen in atopic dermatitis showing a switch from an initial Th2 response to a Th1 response in situ: an immunocytochemical study. *J Allergy Clin Immunol* 97: 828–837

29 Sillevis Smitt JH, Bos JD, Hulsebosch HJ, Krieg SR (1986) *In situ* immunophenotyping of antigen presenting cells and T cell subsets in atopic dermatitis. *Clin Exp Dermatol* 11: 159–168

30 Kägi MK, Wüthrich B, Montano E, Barandun J, Blaser K, Walker Ch (1994) Differential cytokine profiles in peripheral blood lymphocyte supernatants and skin biopsies from patients with different forms of atopic dermatitis, psoriasis and normal individuals. *Int Arch Allergy Immunol* 103: 332–340

31 Bos JD, Van Leent EJM, Sillevis Smitt JH (1998) The millennium criteria for the diagnosis of atopic dermatitis. *Exp Dermatol* 7: 132–138

32 Bos JD (ed) (1997) *Skin immune system (SIS): Cutaneous immunology and clinical immunodermatology.* CRC Press, Boca Raton

Immunological aspects of allergic inflammation: IgE regulation

Gerald R. Dubois, Paul J. Baselmans and Geert C. Mudde

Research Institute Jouveinal, Parke-Davis, 94265 Fresnes, France

Introduction

The increased production of IgE forms the hallmark of the atopic type of allergy and has implications for many cell types involved. The importance of IgE was first recognized in immediate type hypersensitivity or Type I hypersensitivity. The major effector mechanism in this type of hypersensitivity is the IgE-dependent stimulation of tissue mast cells and their circulating equivalent, the basophils. Cross-linking of IgE, bound to the high affinity IgE receptor (FcεRI) expressed on these cells, by allergens results in the rapid release of a variety of mediators, including histamine, leukotrienes, prostaglandins, and several proteases such as tryptase and chymase, and has been demonstrated more recently, interleukin 4 [1–3]. The presence of these mediators results in tissue-specific symptoms like hay-fever, broncho-constriction, and urticaria as classical manifestation of IgE-dependent type I hypersensitivity reactions.

The first evidence that IgE may play a role in antigen capture by antigen presenting cells (APC) was found in patients with atopic dermatitis (AD) [4]. The presence of IgE+ Langerhans cells [5] and CD4+ allergen specific T cells in the skin of many AD patients [6, 7] suggests that presentation of allergens by Langerhans cells to T cells is an important event in the inflammatory skin reaction in these patients. The Fc receptor responsible for the binding of IgE to LC has been identified as the FcεRI [8, 9]. Further evidence for the involvement of FcεRI in IgE-mediated antigen presentation was provided by studies showing that FcεRI expressed on monocytes and blood dendritic cells are able to mediate antigen presentation *via* IgE *in vitro* [10, 11]. In addition, CD23, the low affinity IgE receptor (K_d 10^{-8} M), has also been shown to mediate antigen uptake [12]. CD23 is constitutively expressed on naïve B cells and can be induced on most APC by IL-4.

B cell development

B cells are derived from the bone marrow and their main function is antigen recognition, whereafter they may differentiate into plasma cells. When differentiated into plasma cells, their function changes from antigen recognition to large-scale antibody production. Antigen recognition by B cells occurs *via* interaction with the B cell receptor (BCR). The BCR is an immunoglobulin (Ig) inserted into the B cell surface *via* the addition of a hydrophobic transmembrane region, which enables the Ig to be expressed on the membrane as a receptor. For signaling, the surface Ig (sIg) is complexed with associated membrane proteins. On the basis of their BCR, two different types of mature B cells can be recognized: (1) Naïve B cells, which co-express receptors of the IgM and IgD isotype, and (2) memory B cell, which express a single isotype, either IgG, IgE, or IgA, on their cell surface.

Naïve B cells become immunocompetent after passing through pre-B and immature B cell stages in the bone marrow. During the differentiation of B cells, somatic recombination first creates a functional heavy chain by the recombination of multiple genes, consisting of V (variable), D (diversity), and J (joining) segments and, subsequently, the gene encoding the constant region of IgM (Cμ). The recombined VDJCμ is transcribed, mRNA translated, and the μ-chain is expressed in the cytoplasm of the pre-B cell. The multiple V and J segments in the light chain loci then recombine with the constant region of light chain segments (C_L) to express a light chain that can assemble with μ-chain to form a functional immunoglobulin molecule, in this case IgM. At this stage, the mRNA splicing mechanism favours the recombination of IgM, expressed on the surface (sIgM). Association of sIgM with signaling proteins results in the expression of a functional BCR. During maturation of an immature B cell to a naïve B cell, VDJ now recombines with Cμ and Cδ to form a functional VDJCμCδ gene. This gene is then subsequently spliced into VDJCμ and CDJCδ mRNA, which, after translation, lead to functional heavy chains of respectively IgM and IgD. Association with VJC_L will then result in the expression of functional BCRs of IgM and IgD isotypes respectively. After successful assembly of an IgM receptor, B cells may undergo a process of negative selection and/or receptor editing to eliminate cells with autoreactive receptor specificities [13]. Upon completion of this process, B cells that express both IgM and IgD receptors on their membrane enter the circulation and seed the peripheral lymphoid organs to await an encounter with antigen.

B cell maturation, controlling antigen specificity

The recognition of antigen is the key initial step in the maturation from naïve B cells into memory cells. Antigen receptors on lymphocytes play a central role in immune regulation by transmitting signals that positively or negatively regulate lymphocyte survival, migration, growth, and differentiation (reviewed in [14]). Different anti-gens provoke distinct antibody responses, dependent on concentration of antigen, antigen avidity, and timing and duration of antigen encounter. T cell dependent anti-gens (including allergens) are usually univalent with respect to the B cell receptors, meaning each epitope appears once on a monomeric protein and as a result cannot cross-link a B cell surface receptor (BCR). In contrast, if the B cell encounters the same antigen in multimeric form with the ability to cross-link BCRs, mimicking for example bacterial surfaces, T cell help is not required and B cells are induced to pro-liferate [15]. The lack of BCR cross-linking in T cell dependent responses does not prevent antigen internalization in endosomes which fuse with vesicles containing MHC class II molecules [16]. Processed protein fragments are subsequently folded into the groove of MHC molecules where they are available for recognition by spe-cific Th cells. Regulation of antigen specificity during B cell maturation and termi-nal differentiation is tightly controlled by various mechanisms. The first checkpoint is at the level of naïve B cell activation which requires cognate T cell help by T cells that underwent thymic selection. This ensures that only foreign antigens will induce B cell activation. In the second step, rapid clonal expansion and somatic hypermu-tation in the proliferating, activated B cell population results in low affinity, high affinity and autoreactive mutants. Whereas the low-affinity B cells will die of apop-tosis due to a lack of BCR triggering, high-affinity B cells survive and pick up anti-gen which was present on the FDC for presentation to antigen-specific T cells in the apical light zone of the germinal center. It is assumed that BCRs have to compete with antibody-binding of immune-complexes which are caught and presented by FDC, in order to escape programmed cell death [17]. In the third phase, which involves isotype switching, memory development and plasma cell differentiation, the role of cognate interactions has not been entirely elucidated [18]. CD40-Ligand or antibodies against CD40 and cytokines like IL-4 have the capacity to trigger pro-liferation and class-switch in highly purified B cells regardless of antigen-specificity [19, 20]. This would imply that activated T cells in the outer zone of a germinal cen-ter could randomly induce class-switch in neighboring B cells that survived previous selections. However, there is accumulating evidence that again BCR-mediated sig-naling and antigen presentation will rescue B cells from death by apoptosis in this phase.

One important pair of molecules in the induction of apoptosis is Fas and its counterpart Fas-ligand. Recently, it was postulated that tolerant B cells which have desensitized and downregulated their BCRs, or B cells that express low affinity BCRs fail to reverse the apoptotic signal that is given upon cognate contact with T cells expressing CD40-ligand and Fas-ligand [21]. Fas-ligand is expressed on T cells following TCR signaling and binding of Fas-ligand to Fas induces rapid apoptosis in a variety of cell types including activated B cells [22]. Blockage of the apoptotic cascade *via* signaling of BCRs that interact with antigen rescues B cells, which sub-sequently remain susceptible to proliferation and isotype-switching.

IL-4 and IL-13, key cytokines in IgE regulation

The role of IL-4 in the induction of IgE first became evident in 1986 [23]. Recombinant IL-4 induced substantial amounts of IgE and IgG_1 in lipopolysaccharide (LPS)-stimulated murine B cells, *in vitro*, whereas LPS alone gave rise to only modest amounts of IgG_1 and virtually no IgE [24]. The essential presence of IL-4 for the induction of IgE was further substantiated by the ability of anti-IL-4 antibodies or a monoclonal antibody against the IL-4 receptor (IL-4R) [23, 25, 26] to abrogate IgE production in mice. In addition, a recombinant extracellular domain of the IL-4 receptor inhibited switching to IgE by blocking the IL-4/IL4R interaction [27]. The most compelling evidence for the central role of IL-4 in IgE induction comes from gene targeting experiments. Mice expressing the IL-4 transgene have strikingly high levels of IgE and IgG_1 [28]. Moreover, IL-4 knockout mice were unable to synthesize IgE in response to a parasitic infection [29], re-affirming the pivotal role of this cytokine in the production of IgE in mice.

Evidence for the role of IL-4 in the induction of IgE in humans came from *in vitro* studies using T cell clones [30, 31]. Using a whole range of PMA or anti-CD3 stimulated T clones, a strong correlation was found between helper function of IgE synthesis and the production of IL-4. All T cell clones that were able to induce IgE synthesis in normal B cells were of the Th2 phenotype, meaning that they predominantly produced IL-4. In contrast, T cell clones that had a Th1 phenotype and thus produce IFNγ were not able to induce IgE synthesis in normal B cells. In addition, it was shown that exogenous IL-4 was able to induce IgE synthesis in peripheral blood mononuclear cells. This effect was dose-dependently inhibited by the addition of recombinant IFNγ, indicating the opposite regulatory role of IL-4 and IFNγ in the induction of IgE synthesis *in vitro*. However, using cognate IgE induction high levels of IFNγ failed to inhibit IgE production [32].

More recently, another cytokine, IL-13, has been identified as a switch factor for IgE synthesis in humans [33]. IL-13 has IL-4-like effects on B cells such as induction of CD23 expression, proliferation, and the isotype switching to IgE and IgG4 [34]. In highly purified naïve $sIgD^+$ splenic B cells IgE and IgG4 synthesis can be induced by IL-13 when costimulated with activated T cells, anti-CD40 antibodies, or L cells transfected with human CD40L. Neutralizing antibodies to IL-4 do not affect the induction of IgE and IgG_4 by IL-13, indicating that IL-13 acts in an IL-4-independent way. In general, IL-13 seems to be about two- to five-fold less potent than IL-4 and has neither additive nor synergistic effects with IL-4 in the induction of IgE or IgG_4 synthesis. These observations suggests a common signaling pathway for the induction of IgE by both cytokines. This is further substantiated using an IL-4 mutant protein, in which the tyrosine at the 124 position was changed into an aspartic acid [35]. This mutant retained the ability to bind to the IL-4R but failed to transmit a signal upon receptor binding and was shown to inhibit both IL-4- and IL-13-induced human IgG4 and IgE synthesis and B cell proliferation. Although both human and mouse IL-13 are not active on T cells and exert similar effects on macrophages, they differ in their effect on B cells. Mouse IL-13 was shown to enhances antibody production *in vivo* but evidence for IL-13-induced class switching could not been provided [36]. *In vitro*, IL-13 increased the survival of mouse B cells and hence antibody production [36], but did not increase CD23 production or B cell proliferation.

Both IL-4 and IL-13 are capable of inducing ε-germline transcripts in human B cells, thereby providing the first signal for the switch to IgE synthesis [33]. The molecular mechanisms by which both cytokines induce the germline transcription have recently been delineated. The IL-4 receptor is comprised of two polypeptide chains, an IL-4 binding chain (IL-4R α-chain) and the common cytokine receptor γ-chain, also used by IL-2, IL-7, IL-15 [37–40]. In addition to a ligand-specific IL-13 binding chain, the IL-13 receptor also uses the IL-4R α-chain [34]. Indeed, this common feature results in the activation of similar signal transduction pathways, one of which is the JAK/STAT pathway. This pathway is used by many cytokine receptors and involves the subsequent activation of a family of tyrosine kinases (JAK1, JAK2, JAK3, and TYK2) and signal transducers and activators of transcription (STAT1-6) [41, 42]. Activation *via* IL-4 involves JAK1 and JAK3, associated to respectively the IL-4R α-chain and the γ-chain [43, 44]. Correspondingly, IL-13 activates JAK1 and TYK2 kinases [45]. Both IL-4 and IL13 subsequently signal *via* STAT6 due to docking of STAT6 to the IL-4R α-chain [45, 46]. The region of the germline ε promoter contains a binding site for an IL-4 responsive element [47]. This element is known to bind to STAT6, indicating the involvement of this cascade in ε germline transcription. In addition, STAT6 deficient mice are unable to produce IgE, supporting the hypothesis that STAT6 is critical for class switch [48].

CD40-CD40L interaction

Although both IL-4 and IL13 are sufficient for the initiation of ε germline transcription, additional signals are needed for the expression of mature ε mRNA. The addition of autologous activated T cells can restore the IL-4-dependent IgE synthesis. It is well established that the principle molecules for T cell mediated B cell activation are CD40 and CD40L. The earliest T-B cell interactions during a primary immune response take place outside the lymphoid follicle, where as a result of antigen-specific activation, B cells migrate from the B cell-rich follicles to the T cell rich areas were they encounter antigen-specific T cells [49, 50]. These T cells were previously activated by dendritic cells that invaded the T cell areas of the lymph node after antigen uptake elsewhere in the body. One special feature of dendritic cells is their unique capability to stimulate naïve T cells. This is mainly due to a high expression of co-stimulatory molecules like B7.1 and B7.2 that by triggering CD28 on the T cell are capable of CD40-Ligand upregulation and IL-2 induction in T cells that recognize antigen presented by the APC. In contrast, naïve antigen-presenting B cells which generally express low levels of B7.2, fail to induce sufficient CD28 signaling in resting T cells to provoke a T cell response and as a result, this leads to their deletion or inactivation [51].

CD40 is an integral membrane protein which is expressed by B cells, dendritic cells, follicular dendritic cells, hematopoietic progenitor cells, epithelial cells and carcinomas (reviewed in [19]). The importance of CD40/CD40-Ligand interactions is demonstrated by the fact that patients with X-linked hyper-IgM syndrome due to a mutated CD40L do not develop germinal centers and lack formation of memory cells [52]. There is no class-switch to either IgE-, IgG-, or IgA-producing cells. Cross-linking of CD40 on naïve B cells induces B cell proliferation and addition of cytokines subsequently induce isotype switching. Both IL-4 and IL-13 allow a class-switch to IgE and IgG_4, IL-10 induces IgG_1 and IgG_3 [53], and in the presence of additional TGFβ, IgA_1 and IgA_2 is secreted [54]. Recent evidence by Arpin et al indicates that a prolonged CD40 signaling in isotype-switched B cells will induce formation of memory cells, whereas removal of the signal induces plasma cell differentiation and antibody production [55].

Directing immune responses by antibodies

Besides their function in clearance and killing of pathogens, antibodies have a direct effect on their own regulation and function as a feedback mechanism for B cell activation. In mice, IgM antibodies are able to augment antibody responses to T cell independent antigens such as sheep red blood cells and malaria parasites and this is a complement-dependent phenomenon [56]. The importance of the complement system for a normal antibody response has been demonstrated in humans with complement deficiency (C1-C4) who are unable to produce normal titers of antibodies [57]. Already in 1973, it was hypothesized that complement receptors on B cells could mediate a second signal for B cell activation [58] and to date, the most likely candidate to mediate this effect is CD21 (complement receptor 2) which, in complex with CD19 and TAPA-1, lowers the threshold of BCR stimulation by a factor of 100 [59]. This suggests that for normal antibody responses, complexes of antigen/(IgM) antibody/complement bind both CD21 and BCR to ensure optimal B cell activation.

The reason why allergens preferentially expand Th2-like $CD4^+$ T cells is still unclear. By secretion of Th2 cytokines IL-4, IL-5 and IL-10, they are excellent providers for B cell help in cognate interactions and a possible role of B cells as APC in the polarization into Th2 cells of naïve T cells has been suggested [60,61]. The dominant expression of B7-2 over B7-1 on activated B cells might explain this favorable development of Th2 differentiation [62] combined with the lack of IL-12 production in cognate B-T interactions, which is an important polarizing cytokine for Th1 cells [63, 64]. IgE, which is mainly produced in the Th2 environment, was found to have two important effects on B cells, both mediated by the low-affinity IgE receptor CD23. Human CD23 is expressed in two isoforms; CD23a which is constitutively expressed on naïve B cells and CD23b which is IL-4 inducible on many APC including B cells. In tonsil sections, B cells in the follicular mantle zone are positive for CD23, whereas those in the GC are CD23⁻. In 1989, CD23 was found to mediate antigen presentation *via* IgE on mouse B cells [65] and human Epstein-Barr virus-transformed B cells [66]. A few years later, CD23 was found responsible for enhanced antibody responses which were produced after injection of TNP-specific-IgE in mice receiving antigen-TNP [67].

The proposed mechanism which links these two observations, is that the IgE/CD23/antigen complex is endocytosed by B cells, leading to increased antigen processing and presentation on major histocompatibility complex (MHC) class II molecules to T helper cells. Importantly, the upregulation of the antibody response is specific for the IgE-binding antigen suggesting that again the antibody response appears to be tightly controlled and only antigen-specific B cells are activated. Importantly, B cells cannot use IgG molecules, which implies that in the presence of IgE potentially large numbers of B cells are included in the pool of APC, whereas in the absence of IgE, B cells are excluded from this pool [68]. As a result, it was hypothesized that IgE-mediated antigen presentation might favor B cells above DC as APC for induction of T cell differentiation and thus promote Th2-cell development [69]. This subsequently results in higher antibody production due to cytokine-secretion of IL-4 and IL-10 but also in a chronic induction of IgE-antibodies causing allergy and may lead to allergy expansion.

Not just IgE, but also IgG antibodies can be used by APC for antigen presentation, but in contrast to IgE, IgG antibodies can be positively or negatively involved in the regulation of antigen presentation and antibody responses. Complexes of antigen and IgG are very efficiently taken-up and presented to T cells by APC like monocyte-derived human DC [70].

Two different types of FcγR have been described to mediate antigen presentation; The high affinity FcγRI (CD64) [71] binds both monomeric and complexed IgG, and the low-affinity FcγRII (CD32) [70] which binds complexed IgG only. The isotypes CD32a and CD32bII are both involved in antigen uptake of complexed IgG [72], however CD32bII is not as efficient as the CD32a-form. CD32bI which, together with CD32bII, is the predominantly expressed IgG-receptor type on matured B cells, contains a cytoplasmic inhibitory motif (ITIM: immune-receptor tyrosine-based inhibition motif), which neutralizes signaling of receptors that contain a specific activation motif (ITAM: immune-receptor tyrosine-based activation motif). Prerequisite for this inhibition is that both receptors are triggered at the same time and in close proximity of each other, which occurs when antigen is complexed by antibodies.

Examples for this phenomenon are mixed complexes of IgE and IgG which, in contrast to FcεRI-crosslinking by IgE, prevent mast cell degranulation by co-engagement of FcεRI and CD32bI [73] or the co-engagement of BCR and CD32bI by IgG-bound antigens on B cells [74]. The latter event subsequently abrogates antigen-specific antibody responses in these B cells. Interestingly, IgG_4, the antibody isotype that is co-regulated with IgE, has a relatively low affinity for CD32 compared to IgG_1 and IgG_3, suggesting that this antibody will unlikely interfere with IgE-mediated effector functions on mast cells or B cells. Successful desensitization-therapies however, are generally accompanied by a shift in antigen-specific IgG_4:IgG_1 ratios which ultimately favor IgG_4 [75]. The effects of IgG_4 can therefore not be explained by CD32bI-mediated effects, but might be due to epitope-shielding, interfering or preventing the binding of IgE to allergens [76]. Contrary to the role of IgE in directing Th2 differentiation, the IgG antibody favors monocyte/macrophage-mediated antigen presentation by CD64 and CD32a above B cells as APC, thus directing T cell responses away from the Th2-phenotype [60]. This might explain the success of desensitization-therapies where IgE:IgG ratios favor IgG as allergic disease improves and why addition of allergen-specific IgG antibodies improves clinical signs of atopic dermatitis and allergic bronchial asthma [77, 78].

Concluding remarks

The production of IgE molecules has been the focus of many studies during recent decades. Apart from the "effector" function of IgE, it is now widely accepted that IgE plays an important role in the regulation of the immune response, as has been hypothesized in the early 1990s [79]. The expression on all kinds of APC in humans with or without atopic allergy points to an important role of this molecule which goes further than the defense against parasitic infections. Although beneficial in the short run, eliminating IgE responses in allergic patients may have as yet unclear effects on the ability of these patients to combat natural antigen challenges.

References

1   Arock M, Merle Beral H, Dugas B, Ouaaz F, Le Goff L, Vouldoukis I, Mencia Huerta JM, Schmitt C, Leblond Missenard V, Debre P et al (1993) IL-4 release by human leukemic and activated normal basophils. *J Immunol* 151: 1441–1447

2   Bradding P, Feather IH, Howarth PH, Mueller R, Roberts JA, Britten K, Bews JP, Hunt TC, Okayama Y, Heusser CH et al (1992) Interleukin 4 is localized to and released by human mast cells. *J Exp Med* 176: 1381–1386

3   Brunner T, Heusser CH, Dahinden CA (1993) Human peripheral blood basophils primed by interleukin 3 (IL-3) produce IL-4 in response to immunoglobulin E receptor stimulation. *J Exp Med* 177: 605–611

4   Mudde GC, Van Reijsen FC, Boland GJ, De Gast GC, Bruijnzeel PLB, Bruijnzeel-

Koomen CAFM (1990) Allergen presentation by epidermal Langerhans' cells from patients with atopic dermatitis is mediated by IgE. *Immunology* 69: 335–341

5 Bruijnzeel-Koomen CAFM, van der Donk EM, Bruynzeel PL, Capron M, De Gast GC, Mudde GC (1988) Associated expression of CD1 antigen and Fc receptor for IgE on epidermal Langerhans cells from patients with atopic dermatitis [published erratum appears in *Clin Exp Immunol* 1988 Dec; 74(3): 504]. *Clin Exp Immunol* 74: 137–142

6 Van Reijsen FC, Bruijnzeel-Koomen CAFM, Kalthoff FS, Maggi E, Romagnani S, Westland JKT, Mudde GC (1992) Skin-derived aeroallergen-specific T cell clones of Th2 phenotype in patients with atopic dermatitis. *J Allergy Clin Immunol* 90: 184–193

7 Neumann C, Gutgesell C, Fliegert F, Bonifer R, Herrmann F (1996) Comparative analysis of the frequency of house dust mite specific and nonspecific Th1 and Th2 cells in skin lesions and peripheral blood of patients with atopic dermatitis. *J Mol Med* 74: 401–406

8 Bieber T, de la Salle H, Wollenberg A, Hakimi J, Chizzonite R, Ring J, Hanau D, de la Salle C (1992) Human epidermal Langerhans cells express the high affinity receptor for immunoglobulin E (FcεRI). *J Exp Med* 175: 1285–1290

9 Wang B, Rieger A, Kilgus O, Ochiai K, Maurer D, Fodinger D, Kinet J-P, Stingl G (1992) Epidermal Langerhans cells from normal human skin bind monomeric IgE via FcεRI. *J Exp Med* 175: 1353–1365

10 Maurer D, Ebner C, Reininger B, Fiebiger E, Kraft D, Kinet JP, Stingl G (1995) The high affinity IgE receptor (Fc epsilon RI) mediates IgE-dependent allergen presentation. *J Immunol* 154: 6285–6290

11 Maurer D, Fiebiger S, Ebner C, Reininger B, Fischer GF, Wichlas S, Jouvin MH, Schmitt E, Kraft D, Kinet JP, Stingl G (1996) Peripheral blood dendritic cells express FcεRI as a complex composed of FcεRI α- and FcεRI γ-chains and can use this receptor for IgE-mediated allergen presentation. *J Immunol* 157: 607–616

12 Pirron U, Schlunck T, Prinz JC, Rieber EP (1990) IgE-dependent antigen focussing by human B lymphocytes is mediated by the low-affinity receptor for IgE. *Eur J Immunol* 20: 1547–1551

13 Nemazee D, Buerki K (1989) Clonal deletion of autoreactive B lymphocytes in bone marrow chimeras. *Proc Natl Acad Sci USA* 86: 8039–8043

14 Healy JI, Goodnow CC (1998) Positive versus negative signaling by lymphocyte antigen receptors. *Annu Rev Immunol* 16: 645–670

15 Wortis HH, Teutsch M, Higer M, Zheng J, Parker DC (1995) B-cell activation by crosslinking of surface IgM or ligation of CD40 involves alternative signal pathways and results in different B-cell phenotypes. *Proc Natl Acad Sci USA* 92: 3348–3352

16 Davidson HW, Reid PA, Lanzavecchia A, Watts C (1991) Processed antigen binds to newly synthesized MHC class II molecules in antigen-specific B lymphocytes. *Cell* 67: 105–116

17 Liu YJ, Joshua DE, Williams GT, Smith CA, Gordon J, MacLennan IC (1989) Mechanism of antigen-driven selection in germinal centres. *Nature* 342: 929–931

18 Liu YJ, Malisan F, De Bouteiller O, Guret C, Lebecque S, Bancherau J, Mills FC, Max EE, Martinez Valdez H (1996) Within germinal centers, isotype switching of

immunoglobulin genes occurs after the onset of somatic mutation. *Immunity* 4: 241–250

19   Banchereau J, Bazan F, Blanchard D, Brière F, Galizzi JP, Van Kooten C, Liu YJ, Rousset F, Saeland S (1994) The CD40 antigen and its ligand. *Annu Rev Immunol* 12: 881–922

20   Foy TM, Aruffo A, Bajorath J, Buhlmann JE, Noelle RJ (1996) Immune regulation by CD40 and its ligand GP39. *Annu Rev Immunol* 14: 591–617

21   Rathmell JC, Townsend SE, Xu JCC, Flavell RA, Goodnow CC (1996) Expansion or elimination of B cells *in vivo*: Dual roles for CD40- and Fas (CD95)-ligands modulated by the B cell antigen receptor. *Cell* 87: 319–329

22   Nagata S, Golstein P (1995) The Fas death factor. *Science* 267: 1449–1456

23   Finkelman FD, Katona IM, Urban JF Jr, Snapper CM, Ohara J, Paul WE (1986) Suppression of *in vivo* polyclonal IgE responses by monoclonal antibody to the lymphokine B-cell stimulatory factor 1. *Proc Natl Acad Sci USA* 83: 9675–9678

24   Lebman DA, Coffman RL (1988) Interleukin 4 causes isotype switching to IgE in T cell-stimulated clonal B cell cultures. *J Exp Med* 168: 853–862

25   Finkelman FD, Katona IM, Urban JF Jr, Holmes J, Ohara J, Tung AS, Sample JV, Paul WE (1988) IL-4 is required to generate and sustain *in vivo* IgE responses. *J Immunol* 141: 2335–2341

26   Finkelman FD, Holmes J, Katona IM, Urban JF Jr, Beckmann MP, Park LS, Schooley KA, Coffman RL, Mosmann R (1990) Lymphokine control of *in vivo* immunoglobuline isotype selection. *Annu Rev Immunol* 8: 303–333

27   Garrone P, Djossou O, Galizzi JP, Banchereau J (1991) A recombinant extracellular domain of the human interleukin 4 receptor inhibits the biological effects of interleukin 4 on T and B lymphocytes. *Eur J Immunol* 21: 1365–1369

28   Tepper RI, Levinson DA, Stanger BZ, Campos Torres J, Abbas AK, Leder P (1990) IL-4 induces allergic-like inflammatory disease and alters T cell development in transgenic mice. *Cell* 62: 457–467

29   Kuhn R, Rajewsky K, Muller W (1991) Generation and analysis of interleukin-4 deficient mice. *Science* 254: 707–710

30   Del Prete GF, Maggi E, Parronchi P, Chretien I, Jiri A, Macchia D, Ricci M, Banchereau J, De Vries J (1988) IL-4 is an essential factor for IgE synthesis induced *in vitro* by human T cell clones and their supernatants. *J Immunol* 140: 4193–4199

31   Pene J, Rousset F, Briere F, Chretien I, Paliard X, Banchereau J, Spits H, de Vries JE (1988) IgE production by normal human B cells induced by alloreactive T cell clones is mediated by IL-4 and suppressed by IFN-gamma. *J Immunol* 141: 1218–1224

32   Armerding D, Van Reijsen FC, Hren A, Mudde GC (1993) Induction of IgE and IgG1 in human B cell cultures with staphylococcal superantigens: Role of helper T cell interaction, resistance to interferon-gamma. *Immunobiology* 188: 259–273

33   Punnonen J, Aversa G, Cocks BG, McKenzie AN, Menon S, Zurawski G, de Waal Malefyt R, de Vries JE (1993) Interleukin 13 induces interleukin 4-independent IgG4 and IgE

synthesis and CD23 expression by human B cells. *Proc Natl Acad Sci USA* 90: 3730–3734

34   Zurawski G, de Vries JE (1994) Interleukin 13, an interleukin 4-like cytokine that acts on monocytes and B cells, but not on T cells. *Immunol Today* 15: 19–26

35   Aversa G, Punnonen J, Cocks BG, de Waal Malefyt R, Vega F Jr, Zurawski SM, Zurawski G, de Vries JE (1993) An interleukin 4 (IL-4) mutant protein inhibits both IL-4 or IL-13-induced human immunoglobulin G4 (IgG4) and IgE synthesis and B cell proliferation: support for a common component shared by IL-4 and IL-13 receptors. *J Exp Med* 178: 2213–2218

36   Lai YH, Mosmann RL (1999) Mouse IL-13 enhances antibody production *in vivo* and acts directly on B cells *in vitro* to increase survival and hence antibody production. *J Immunol* 162: 78–87

37   Giri JG, Ahdieh M, Eisenman J, Shanebeck K, Grabstein K, Kumaki S, Namen A, Park LS, Cosman D, Anderson D (1994) Utilization of the b and g chain of the IL-2 receptor by the novel cytokine IL-15. *EMBO J* 13: 2822–2830

38   Kondo M, Takeshita T, Ishii N, Nakamura M, Watanabe S, Arai K, Sugamura K (1993) Sharing of the interleukin-2 (IL-2) receptor gamma chain between receptors for IL-2 and IL-4 [see comments]. *Science* 262: 1874–1877

39   Kondo M, Takeshita T, Higuchi M, Nakamura M, Sudo T, Nishikawa S, Sugamura K (1994) Functional participation of the IL-2 receptor γ-chain in IL-7 receptor complexes. *Science* 263: 1453–1456

40   Noguchi M, Nakamura Y, Russell SM, Ziegler SF, Tsang M, Cao X, Leonard WJ (1993) Interleukin-2 receptor γ-chain: A functional component of the interleukin-7 receptor. *Science* 262: 1877–1881

41   Ihle JN, Witthuhn BA, Quelle FW, Yamamoto K, Theirfelder WE, Kreider B, Silvennoinen O (1994) Signaling by the cytokine receptor superfamily: JAKs and STATs. *Trends Biochem Sci* 19: 222–227

42   Ziemiecki A, Harpur A, Wilks AF (1994) JAK protein tyrosine kinases: their role in cytokine signaling. *Trends Cell Biol* 4: 207–212

43   Yin T, Tsang ML-S, Yang Y-C (1994) JAK1 kinase forms complexes with interleukin-4 receptor and 4PS/IRS-1-like protein and is activated by IL-4 and IL-7 in T lymphocytes. *J Biol Chem* 269: 26614–26617

44   Russell SM, Johnston JA, Noguchi M, Kawamura M, Bacon CM, Friedmann MC, Berg M, McVicar DW, Witthuhn BA, Silvennoinen O, Goldman AS, Schmalstieg FC, Ihle JN, O'Shea JJ, Leonard WJ (1994) Interaction of IL-2Rβ and gc chains with jak1 and jak3: inplications for XSCID and XCID. *Science* 266: 1042–1045

45   Welham MJ, Learmonth L, Bone H, Schrader JW (1995) Interleukin-13 signal transduction in lymphohemopoetic cells. *J Biol Chem* 270: 12286–12296

46   Hou J, Schindler U, Henzel WJ, Ho TC, Brasseur M, McKnight SL (1994) An interleukin-4-induced transcription factor: IL-4 Stat. *Science* 265: 1701–1706

47   Kohler I, Alliger P, Minty A, Caput D, Ferrara P, Holl-Neugebauer B, Rank G, Rieber EP (1994) Human interleukin-13 activates the interleukin-4-dependent transcription

factor NF-IL4 sharing a DNA binding motif with an interferon-gamma-induced nuclear binding factor. *FEBS Lett* 345: 187–192

48 Shimoda K, Van Deursen J, Sangster MY, Sarawar SR, Carson RT, Tripp RA, Chu C, Quelle FW, Nosaka T, Vignali DAA, Doherty PC, Grosveld G, Paul WE, Ihle JN (1996) Lack of IL-4-induced Th2 response and IgE class switching in mice with disrupted *Stat6* gene. *Nature* 380: 630–633

49 MacLennan IC, Gulbranson-Judge A, Toellner KM, Casamayor-Palleja M, Chan E, Sze DM, Luther SA, Orbea HA (1997) The changing preference of T and B cells for partners as T-dependent antibody responses develop. *Immunological Reviews* 156: 53–66

50 Garside P, Ingulli E, Merica RR, Johnson JG, Noelle RJ, Jenkins MK (1998) Visualization of specific B and T lymphocyte interactions in the lymph node. *Science* 281: 96–99

51 Fuchs EJ, Matzinger P (1992) B cells turn off virgin but not memory T cells. *Science* 258: 1156–1159

52 Facchetti F, Appiani C, Salvi L, Levy J, Notarangelo LD (1995) Immunohistologic analysis of ineffective CD40-CD40 ligand interaction in lymphoid tissues from patients with X-linked immunodeficiency with hyper-IgM: Abortive germinal center cell reaction and severe depletion of follicular dendritic cells. *J Immunol* 154: 6624–6633

53 Brière F, Servet-Delprat C, Bridon J-M, Saint-Remy J-MR, Banchereau J (1994) Human interleukin 10 induces naive surface immunoglobulin D+ (sIgD+) B cells to secrete IgG1 and IgG3. *J Exp Med* 179: 757–762

54 Zan H, Cerutti A, Dramitinos P, Schaffer A, Casali P (1998) CD40 engagement triggers switching to IgA1 and IgA2 in human B cells through induction of endogenous TGF-β: Evidence for TGF-β but not IL-10-Dependent direct sμ-->Sα and sequential sμ-->Sgamma, sgamma-->Sα DNA recombination. *J Immunol* 161: 5217–5225

55 Arpin C, Déchanet J, Van Kooten C, Merville P, Grouard G, Brière F, Banchereau J, Liu YJ (1995) Generation of memory B cells and plasma cells *in vitro*. *Science* 268: 720–722

56 Heyman B, Pilstrom L, Shulman MJ (1988) Complement activation is required for IgM-mediated enhancement of the antibody response. *J Exp Med* 167: 1999–2004

57 Jackson CG, Ochs HD, Wedgwood RJ (1979) Immune response of a patient with deficiency of the fourth component of complement and systemic lupus erythematosus. *N Engl J Med* 300: 1124–1129

58 Dukor P, Hartmann KU (1973) Hypothesis. Bound C3 as the second signal for B-cell activation. *Cell Immunol* 7: 349–356

59 Carter RH, Fearon DT (1992) CD19: lowering the threshold for antigen receptor stimulation of B lymphocytes. *Science* 256: 105–107

60 Gajewski TF, Pinnas M, Wong T, Fitch FW (1991) Murine Th1 and Th2 clones proliferate optimally in response to distinct antigen-presenting cell populations. *J Immunol* 146: 1750–1758

61 Gajewski TF, Lancki DW, Stack R, Fitch FW (1994) "Anergy" of T_H0 helper T lymphocytes induces downregulation of T_H1 characteristics and a transition to a T_H2-like phenotype. *J Exp Med* 179: 481–491

62 Kuchroo VK, Das MP, Brown JA, Ranger AM, Zamvil SS, Sobel RA, Weiner HL,

Nabavi N, Glimcher LH (1995) B7-1 and B7-2 costimulatory molecules activate differentially the Th1/Th2 developmental pathways: Application to autoimmune disease therapy. *Cell* 80: 707–718

63 Guéry JC, Ria F, Galbiati F, Adorini L (1997) Normal B cells fail to secrete interleukin-12. *Eur J Immunol* 27: 1632–1639

64 Maruo S, Oh-hora M, Ahn HJ, Ono S, Wysocka M, Kaneko Y, Yagita H, Okumura K, Kikutani H, Kishimoto T, Kobayashi M, Hamaoka T, Trinchieri G, Fujiwara H (1997) B cells regulate CD40 ligand-induced IL-12 production in antigen-presenting cells (APC) during T cell/APC interactions. *J Immunol* 158: 120–126

65 Kehry MR, Yamashita LC (1989) Low-affinity IgE receptor (CD23) function on mouse B cells: Role in IgE-dependent antigen focusing. *Proc Natl Acad Sci USA* 86: 7556–7560

66 Pirron U, Schlunck T, Prinz JC, Rieber EP (1990) IgE-dependent antigen focussing by human B lymphocytes is mediated by the low-affinity receptor for IgE. *Eur J Immunol* 20: 1547–1551

67 Gustavsson S, Hjulström S, Tianmin L, Heyman B (1994) CD23/IgE-mediated regulation of the specific antibody response *in vivo*. *J Immunol* 152: 4793–4800

68 Bheekha Escura R, Wasserbauer E, Hammerschmid F, Pearce A, Kidd P, Mudde GC (1995) Regulation and targetting of T-cell immune responses by IgE and IgG antibodies. *Immunology* 86: 343–350

69 Mudde GC, Bheekha R, Bruijnzeel-Koomen CAFM (1995) Consequences for IgE/CD23 mediated antigen presentation in allergy. *Immunol Today* 16: 380–383

70 Sallusto F, Lanzavecchia A (1994) Efficient presentation of soluble antigen by cultured human dendritic cells is maintained by granulocyte/macrophage colony-stimulating factor plus interleukin 4 and downregulated by tumor necrosis factor α. *J Exp Med* 179: 1109–1118

71 Gosselin EJ, Wardwell K, Gosselin DR, Alter N, Fisher JL, Guyre PM (1992) Enhanced antigen presentation using human Fcgamma receptor (monocyte/macrophage)-specific immunogens. *J Immunol* 149: 3477–3481

72 Van Den Herik-Oudijk IE, Westerdaal NAC, Henriquez NV, Capel PJA, Van de Winkel JGJ (1994) Functional analysis of human FcgammaRII (CD32) isoforms expressed in B lymphocytes. *J Immunol* 152: 574–585

73 Daëron M, Malbec O, Latour S, Arock M, Fridman WH (1995) Regulation of high-affinity IgE receptor-mediated mast cell activation by murine low-affinity IgG receptors. *J Clin Invest* 95: 577–585

74 Amigorena S, Bonnerot C, Drake JR, Choquet D, Hunziker W, Guillet J-G, Webster P, Sautes C, Mellman I, Fridman WH (1992) Cytoplasmic domain heterogeneity and functions of IgG Fc receptors in B lymphocytes. *Science* 256: 1808–1812

75 Aalberse RC, van der Gaag R, van Leeuwen J (1983) Serologic aspects of IgG4 antibodies. I. Prolonged immunization results in an IgG4-restricted response. *J Immunol* 130: 722–726

76 Hussain R, Poindexter RW, Ottesen EA (1992) Control of allergic reactivity in human

filariasis. Predominant localization of blocking antibody to the IgG4 subclass. *J Immunol* 148: 2731–2737

77 Jacquemin MG, Machiels JJ, Lebrun PM, Saint-Remy J-MR (1990) Successful treatment of atopic dermatitis with complexes of allergen and specific antibodies. *Lancet* 335: 1468–1469

78 Machiels JJ, Somville MA, Lebrun PM, Lebecque SJ, Jacquemin MG, Saint-Remy J-MR (1990) Allergic bronchial asthma due to Dermatophagoides pteronyssinus hypersensitivity can be efficiently treated by inoculation of allergen-antibody complexes. *J Clin Invest* 85: 1024–1035

79 Mudde GC, Hansel TT, Van Reijsen FC, Osterhoff BF, Bruijnzeel-Koomen CAFM (1990) IgE: An immunoglobulin specialized in antigen capture. *Immunol Today* 11: 440–443

The role of T cells in the pathogenesis of atopic dermatitis

Els Van Hoffen and Frank C. Van Reijsen

Department of Dermatology/Allergology, University Medical Center Utrecht, P.O. Box 85.500, 3508 GA Utrecht, The Netherlands

Introduction

T cells play a central role in the activation and regulation of immune responses. The T cell recognizes antigen, *via* the interaction between a MHC-peptide complex on the antigen presenting cells (APC) and the T cell receptor (TCR) on the T cells, leading to the induction of cytokine production by the T cell. The repertoire of secreted cytokines has an evident influence on the resulting response, directing towards either a cell-mediated T helper 1 (Th1) response or towards a humoral Th2 response [1, 2]. This will have a pronounced effect on the development of different forms of diseases. Atopic dermatitis (AD) is one of the diseases that may result from an unbalanced response of T cells upon allergen-recognition.

T cells have been demonstrated to be present in lesional skin of patients with AD [3]. This chapter will focus on the role of these T cells in the pathogenesis of AD. The exact mechanism of the induction and maintenance of AD lesions is not yet fully understood. A good model to study the induction of eczematous skin lesions is the atopy patch test (APT) [4, 5]. In the APT, allergens are applied onto the skin, by epicutaneous occlusion. An eczematous reaction is observed after 24 h, which declines after 72–96 h. The APT reaction at 24 h closely resembles acute AD lesions, whereas at 48–72 h, the lesions resemble the chronic form of AD [4, 6]. A positive APT reaction occurs in about 50% of the patients with AD with specific serum-IgE for house-dust mite or grass pollen, and not in atopic patients without AD [5]. The APT has been used to study, for example, cellular infiltration, antigen presentation, and cytokine production in the skin of AD patients.

The available data about T cell activation, cytokine production, T cell homing, and T cell memory are reviewed in this chapter. At the end of the chapter, the data will be summarized in a model which presents our current view on the contribution of T cells to AD.

T cell activation in the skin

For the activation of T cells, antigen needs to be processed and presented as peptides by APC. The most important APC in the skin are the Langerhans cells (LC), present in the epidermis. The number of LC is increased in lesional skin of AD patients, as compared to normal and non-lesional skin [3]. Already several years ago, it was observed that LC in AD skin can efficiently bind IgE [7, 8]; this implies that they may use antigen-specific IgE to capture antigen, in the same way as has been described for B cells [9]. It was shown that LC express both the high and low affinity receptor for IgE (FcεRI and FcεRII or CD23, respectively) [10, 11]. Although both receptors are expressed on the LC, IgE binds mainly to the high affinity IgE receptor [7, 12].

A positive APT reaction was shown to correlate with the serum level of antigen-specific IgE, and also with the presence of IgE-bearing LC in lesional skin [4,13]. These data may indicate that LC pick up IgE samples from the serum; this IgE, bound to the FcεRI, is then used to capture allergens from the environment, which, in the case of the APT, are applied epicutaneously. The captured allergen can be internalized, processed and presented as peptides to the T cells. *In vitro*, it was demonstrated that LC are only able to activate house-dust mite-specific T cells if the LC bear antigen-specific IgE [14]. When the LC are triggered by the captured allergen, they may leave the skin, migrate to the regional lymph nodes and present the antigens to the T cells there [15, 16]. However, the demonstration of both IgE+ LC and T cells in lesional as well as non-lesional skin [17], may imply that antigen presentation can also take place locally, in the skin itself. This would be in accordance with the observation that the APT is positive in most AD patients with IgE+ LC, where the LC can directly induce T cell activation, whereas the APT is negative in most AD patients with IgE− LC, in which the LC are not able to provoke local T cell activation, due to lack of proper allergen-uptake and presentation in the skin [17].

Together, these data indicate that activation of T cells in AD not necessarily occurs in lymph nodes, but may very well be achieved by direct antigen presentation in the skin itself.

Phenotype and memory of T cells in the skin

The T cell population infiltrating the eczematous skin is not a homogeneous population. The different T cell subsets present in the skin have been analyzed using immunohistochemistry and T cell culture. The subsets can be discriminated based on the surface expression of certain markers, but also on the expression of different cytokine profiles. Although the presence of T cells in the skin is dynamic, with new T cells entering and others leaving the skin, T cells can also remain in the skin for at least several years to retain memory towards the allergen.

T cell subsets defined by surface markers

T cells are already present in normal human skin, with the majority of the cells being present in the dermis. The CD4:CD8 ratio in normal skin is about 1:1, or is biased towards the CD4+ subset [18]. In the dermis of acute eczematous skin, there is an influx of T cells, with a preference for the CD4+ subset [19]. This correlates with the phenotype of T cells that can be cultured from skin biopsies, of which the majority is CD4+ [20, 21]. The number of T cells in the epidermis is also increased. It was recently shown that the extent of CD4+ T cell infiltration in the skin correlates with the level of interleukin-16 (IL-16) mRNA expression [22]. IL-16 can be produced by, among others, eosinophils, mast cells and T cells [23, 24]. IL-16 is a cytokine that acts as a chemoattractant for CD4+ T cells, *via* interaction with the CD4 molecule [25]. Thereby, IL-16 may be involved in the accumulation of CD4+ T cells in acute AD lesions. In chronic lesional skin, the T cell number in the epidermis is strongly reduced [3, 19, 26]. In the dermis, however, an infiltrate of T cells and macrophages stays present [3, 19, 27].

Most of the T cells present in the skin are in an activated state, expressing HLA-DR, the interleukin-2 receptor CD25, or CD45RO [19, 28–30]. This activated phenotype is also observed in the population of allergen-specific peripheral blood T cells, which express the skin-homing molecule, cutaneous lymphocyte-associated antigen (CLA), and which may thereby preferentially migrate to the skin upon an allergen challenge [31]. *In vitro*, IL-16 in synergy with IL-2 was shown to induce CD4+ T cell activation, measured by the expression of CD25 and CD45RO [32]. In the skin, IL-16 may have similar effects, not only attracting CD4+ T cells, but also enhancing their activation.

T cell subsets defined by cytokine expression

CD4+ T cells, the main source of cytokines, can be subdivided into two functionally distinct subsets, based on the profile of cytokines they produce. In mice, T helper 1 (Th1) cells mainly produce IL-2 and interferon γ (IFNγ), whereas Th2 cells produce IL-4, IL-5 and IL-10 [1, 2]. In humans, the restricted expression of cytokines by either T cell subtype is less strict, with only IFNγ and IL-4 as the main characteristic cytokines of the Th1 and Th2 cells, respectively [33]. The main driving forces in the development of naïve T cells towards a Th1 or Th2 phenotype are the cytokines IL-12 and IL-4, respectively; addition of IL-12 and anti-IL-4 antibodies to naïve T cells during activation results in Th1, whereas addition of IL-4 and anti-IL-12 antibodies causes Th2 cell development [34, 35].

Th2 cells induce humoral immunity. The production of, for example, IL-4 and IL-13 by Th2 cells results in the isotype switch to IgE in B cells [36]. Because induction of allergen-specific IgE plays a role in atopic diseases, several studies analyzed the expression of Th1 and Th2 cytokines in these diseases. In peripheral blood, stimulation of mononuclear cells from AD patients indeed displays a reduced production of IFNγ versus an increased production of IL-4 [37, 38], pointing towards a disbalance in the production of these cytokines *in vivo*. This Th2 bias was also demonstrated in the APT, as a model for AD (Fig. 1). During the early response induced by the allergen, IL-4 expression was shown to predominate over the expression of IFNγ. However, the reverse was true in the late phase of the response, where IFNγ expression was increased and IL-4 expression had decreased, resulting in a switch to a Th0/Th1 mediated chronic response [6]. The cytokine expression in the late phase of the APT closely resembled the expression observed in chronic lesional skin of AD patients. This shift from a Th2 to a Th1 response was demonstrated in other studies as well. When acute AD skin lesions were compared to chronic lesions, not only IL-4, but also IL-5 and IL-13 expression was higher in acute lesions, whereas IFNγ and IL-12 expression was higher in chronic lesions [39–41]. Also *in vitro*, T cell clones, propagated from chronic skin lesions, mainly expressed IFNγ, only some of which expressed IFNγ in combination with IL-4 [42]. Another indication that IFNγ may be involved in the "chronification" of the allergic response was the observation that the expression of IFNγ dropped to levels observed in non-involved skin after successful treatment of eczematous skin [43].

From the different studies, a model could be postulated, as reviewed in [44]. The expression of IL-4 and IL-5 in acute skin lesions may cause the early influx of eosinophils, as is observed early in the APT. *In vitro*, eosinophils were shown to produce IL-12 [45]. In addition, monocytes, macrophages and LC can be induced to produce IL-12. This production of IL-12 by APC can even be induced by Th2 cells *in vitro* [46]. IL-12, demonstrated to be up-regulated in chronic skin, may contribute to the switch from the acute Th2-mediated response with IL-4 and IL-5 production to the continued Th1-mediated chronic response, with a predominance of IFNγ production.

T cell memory in the skin

As mentioned before, T cells in AD lesions can express CD45RO, a marker of activated/memory T cells. That T cell memory is present in the skin was shown by Bohle et al. [47]. In that study, T cell clones, isolated from APT skin in 1990 and 1991 were also present in T cell lines obtained in 1993, and in lesional AD skin in 1990, all obtained from the same AD patient. Analysis of T cell lines and clones, cultured from skin, demonstrated that only part of the T cells is specific for the allergen that elicited the skin reaction [42]. These T cells may be the memory T cells. The other T cells are bystander T cells with other specificities. These cells may become activated secondarily, by the action of the previously activated memory T cells. This phenomenon has been shown before *in vitro* [48]. This secondary acti-

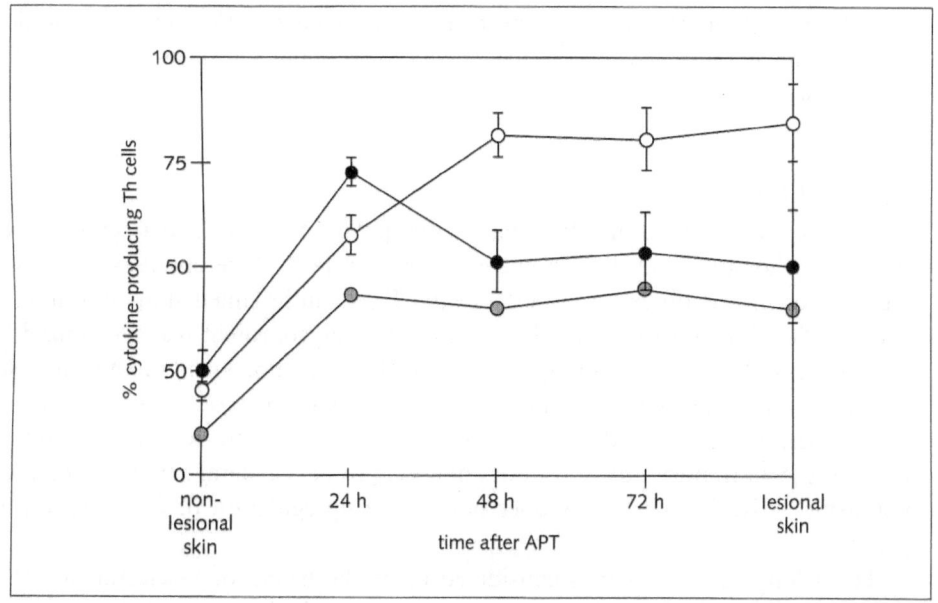

Figure 1
Cytokine expression in skin of patients with atopic dermatitis.
Using immunohistochemistry, the percentage of infiltrating T cells expressing IL-4 and IFNγ was measured in non-lesional skin, lesional skin, and at various time points after an atopy patch test (APT) with house-dust mite. Black circles indicate the percentage of IL-4-expressing cells (both IL-4 single- and IL-4/IFNγ double-positive). The white circles indicate the percentage of IFNγ-expressing cells (both IFNγ single- and IL-4/IFNγ double-positive). The gray circles indicate the IL-4/IFNγ double-positive cells. A shift from a Th2 (IL-4 > IFNγ) towards a Th0/Th1 response (IFNγ ≥ IL-4) can be observed during time after the APT. The cytokine expression at 72 h correlates well with the expression observed in lesional skin.

vation of bystander T cells may also contribute to the Th2 to Th1 shift in cytokine production. The allergen-specific T cells, primarily present and producing cytokines, may be more Th2 prone than the bystander T cells, which are activated later.

Homing of T cells to the skin

A poorly understood aspect of atopic diseases is the localization: why has one individual an allergy to house-dust mite, and suffers from allergic asthma, whereas another individual, also with allergy to house-dust mite, suffers from atopic dermatitis? This paragraph will describe current ideas about the specific homing of T cells to the skin.

CLA and E-selectin

It was already suggested in 1980 that a subpopulation of T cells may exist, which specifically homes to the skin upon certain triggers [49]. Preferential expression of a glycoprotein was observed on 85% of the T cells in inflamed skin, and on only about 15% of peripheral blood T cells [50]. This glycoprotein was recognized by the HECA-452 rat IgM antibody, and was called the cutaneous lymphocyte-associated antigen (CLA). CLA+ peripheral blood T cells seemed to belong to the memory subset, with high adhesion molecule expression and low expression of CD45RA [50]. Importantly, CLA does not seem to be a general activation marker; mitogenic activation of T cells does not induce upregulation of CLA expression [51].

For a long time, CLA was considered to be the ligand of E-selectin, which is expressed on, among others, vascular endothelium [52]. Recently, CLA was identified as a modified form of P-selectin glycoprotein ligand-1 (PSGL-1) [53]. PSGL-1 is constitutively expressed on all T cells [54]. Post-transcriptional glycosylation of PSGL-1 by fucosyltransferase VII results in the expression of CLA, as recognized by the HECA-452 antibody [53]. This modification of PSGL-1 has no major effect on the capacity of binding to P-selectin; CLA+ and CLA− cells can adhere equally to P-selectin. In contrast, CLA expression does strongly improve binding to E-selectin, as has also been demonstrated earlier [53, 55]. This may explain why CLA has always been considered to be a ligand of E-selectin.

Although CLA is considered to be a skin-homing molecule, the selectins to which CLA can bind are not expressed selectively on skin endothelium. Lung tissue upregulates E-selectin expression after allergen exposure [56]. E-selectin expression has also been observed in liver and kidney during inflammation [57, 58]. This widespread tissue distribution of selectins implies that the T cells themselves, more than the expression of selectins in the skin, determine the selective homing of T cells to the skin.

CLA as a specific T cell skin homing molecule

Several studies demonstrated that CLA may be involved in specific homing of T cells to the skin, not only in AD but also in other skin disorders. In cutaneous T cell lymphomas, the extent of skin involvement, but not the involvement of lymph nodes, correlated with the level of CLA expression in peripheral blood [59]. Analysis of CLA expression in patients with psoriatic arthritis showed a significantly higher percentage of CLA-expressing cells in psoriatic skin than in psoriatic synovium [60]. Also in allergic cutaneous drug reactions, an association with CLA expression was observed; the number of CLA+ T cells in the peripheral blood increased during the drug reaction, and decreased with the disappearance of skin symptoms [61]. In patients with cow's milk-induced problems, the number of peripheral blood CLA+ T cells after casein stimulation was significantly increased in patients with cow's milk-induced AD compared to patients with cow's milk induced enterocolitis or to non-atopic controls [62]. An interesting observation was that T cells responding to house-dust mite in patients with AD resided mainly in the CLA+ T cell subset, whereas this response in patients with atopic asthma was mainly present in the CLA− T cell subset [63]. Additionally, in patients with both asthma and AD, proliferating T cells were observed in both the CLA+ and the CLA− subset, suggesting that the CLA+ cells are involved in the skin problems, and the CLA− cells in the lung problems of these patients [64].

The picture arising from these studies seems quite convincing: CLA expression identifies those T cells, which migrate to the skin after a proper stimulus. However, it appears that this picture is too simple. CLA was shown not to be absolutely required for migration into the skin. In the APT reaction, the number of T cells in the skin increased significantly upon allergen application. However, the number of CLA+ T cells within the total T cell population remained constant, implying that most of the T cells entering the skin were CLA− [27]. Patients with an inborn defect in fucose metabolism are not able to express CLA. Despite this defect, however, KLH injection could still induce the influx of a large number of T cells into the skin. These T cells were all CLA− [65]. Although the infiltrating T cells did not behave entirely normal, since redness and swelling of the skin, as signs of delayed-type hypersensitivity, were not observed, the CLA− T cells were able to migrate into the skin.

CLA as a specific homing molecule for T cells is also undermined. CD34+ precursors of LC in the peripheral blood were also shown to be distinguishable from dendritic cells (DC) by their expression of CLA. The CLA+ CD34+ precursor cells developed, after culture, into LC with the characteristic Birbeck granules, whereas the CLA− CD34+ precursor cells developed into DC without Birbeck granules [66]. In addition, CD1a+ LC in diseased skin have an increased expression of CLA compared to normal skin [67]. This may imply that CLA is also involved in the preferential homing of LC to the skin.

The above-mentioned data indicate that neither selectin expression nor CLA expression per se is sufficient for the skin-homing properties of T cells. It may be more reasonable that a combination of events, such as antigen-specific triggering, selectin and CLA expression, at the right time and the right place, is required to get T cells homing to the skin. In addition, CLA should be considered as a molecule more generally involved in homing, not just of T cells, but also of other cell types, which specifically migrate into the skin.

Chemokines and their receptors

In addition to CLA and E-selectin, other factors are involved in the homing and migration of T cells. T cells can be attracted towards the site of inflammation by chemokines, which act *via* chemokine receptors on the surface of the T cells. The best described chemokines that are thought to play a role in the migration of T cells in allergic responses are eotaxin and RANTES (regulated upon activation, normal T cell expressed and secreted). Both chemokines can bind to the chemokine receptor CCR3 [68]. CCR3 is expressed on eosinophils, activated/ memory T cells and basophils [68]. On T cells, CCR3 was described to be preferentially expressed on the Th2 subset [69, 70]. Eotaxin can only act *via* CCR3, whereas RANTES can act *via* more receptors, such as CCR1, CCR3 and CCR5 [68, 71]. Eotaxin can be produced by eosinophils, but can also be induced by IL-4 *in vitro* in skin fibroblasts [72, 73]. RANTES can be produced *in vitro* by keratinocytes [74]. Both chemokines were shown to induce T cell migration *in vitro* [75, 76]. In response to eotaxin, a preferential migration was shown of CCR3-expressing T cells *in vitro* [76]. *In vivo*, these T cells were only observed when eosinophils were present in the tissue [77]. This may imply that production of eotaxin by eosinophils, in addition to eotaxin production by skin fibroblasts, attracts T cells of the Th2 phenotype. Because eosinophils themselves also express the CCR3, they can also migrate in response to eotaxin and RANTES [72, 78]. Intradermal injection of RANTES, which was demonstrated to induce migration of eosinophils and activated T cells, also resulted in the induction of E-selectin [78]. Combining these data with the concept of homing of CLA[+] T cells *via* E-selectin, this again may demonstrate that homing is not only a matter of CLA expression, but involves other factors as well, one of which may be the induction of E-selectin expression by RANTES.

Superantigens

Bacterial colonization is a common observation in AD patients. *Staphylococcus aureus* (*S. aureus*) for example, could be isolated from lesional AD skin in more than 80% of AD patients [79]. That bacterial infection can enhance the severity of AD was shown by the fact that a better improvement of AD was seen in patients treated with a combination of corticosteroids and antibiotics than in patients treated with corticosteroids alone [80]. The effect of bacterial infections on AD is thought to be mainly mediated by bacterial products, especially the bacterial superantigens. Superantigens are potent, antigen-independent, T cell-activating molecules. *Via* direct interaction with the MHC molecule and the TCR, superantigens can stimulate 20–25% of resting T cells [81]. Different superantigens can preferentially activate T cells expressing certain TCR Vβ-chain families. For example, staphylococcal enterotoxin B (SEB), derived from S. aureus, preferentially activates T cells expressing Vβ 3, 12, 14, 15, 17 and 20 [82].

The T cell activating capacity of superantigens may explain the effect of bacterial infection on AD. Direct evidence was given by application of SEB onto normal and intact AD skin, which was shown to induce skin lesions very similar to dermatitis, with erythema and induration [83]. Indirect indications are that T cells from children with active AD show an increased number of T cells expressing the superantigen-related TCR Vβ2 and Vβ5 in the CLA+ but not in the CLA− T cell population. The CLA+ T cells in these infants were more activated as well, with an increased HLA-DR expression, than in children with inactive AD or in healthy controls [84]. Superantigen derived from S. aureus cultured from AD skin was able to induce TCR Vβ skewing in the CD4+ and CD8+ CLA+ T cells of patients with AD, but not in patients with plaque psoriasis or in healthy individuals [85]. In vitro, SEB induced an increased level of IL-4, IL-5 and IgE production versus a decreased IFNγ production in PBMC from patients with AD compared to healthy controls [86]. In children with AD, a reduction in both the number of IFNγ-producing cells and in the level of IFNγ production per cell was observed, in CD4+ and CD8+ T cells and in NK cells in peripheral blood [87]. In these children, a correlation was also described with sensitization to superantigens. The severity of AD, and also the levels of specific IgE to different allergens were higher than in non-superantigen-sensitized children. IgE levels to superantigens were even better correlated with disease severity than total IgE levels [88]. Staphylococcal toxic shock syndrome toxin-1 (TSST-1) was shown to induce IgE production in PBMC of AD patients. This IgE induction was, in the pollen season, even independent of the addition of exogenous IL-4. Together with this increased IgE synthesis, a reduced IFNγ production was observed. The increase in IgE production was shown to result from allergen-specific IgE synthesis [89]. These data, however, are in contrast to other observations. In PBMC from atopic individuals, not an increase but a reduction of IL-4 and IgE production was demonstrated in response to SEB and SEA [90]. Also, in a murine humanized SCID model, exposition of the skin to a combination of SEB and the house dust mite allergen Der p, which induced a strong skin-inflammatory response, decreased the production of IgE [91]. Exposition of monocytes to superantigen was shown to provoke IFNγ production [92]. If this induction of IFNγ would also occur in vivo, it may enhance the shift from an acute Th2 to a chronic Th1 mediated response in AD.

These contradictory data may result from differential effects of superantigens. In any case, superantigens are shown to be able to have a pronounced effect on the induction of acute AD lesions, on T cell activation and on the induction of different types of cytokines in various cell types. Together, these effects may explain why bacterial infections could affect the severity and the course of both the acute and chronic inflammatory response in AD.

T cells in the pathogenesis of AD: a model

In this last paragraph, the data reviewed in the previous paragraphs will be combined in a chronologic model, which summarizes the possible role of T cells in AD. Part of the model is illustrated in Figure 2; numbers in the text between parentheses refer to the numbers in Figure 2.

Allergen may enter the body *via* different routes. Inhaled and ingested allergen may enter the circulation. *Via* the circulation, they can reach the lymph nodes, where they can be picked up by follicular dendritic cells (FDC) and B cells. The FDC present the allergen to B cells, the B cells present it to T cells, and the T cells in turn can induce IgE production by B cells. Allergen can be retained on FDC for a long time [93], enabling a continuous induction of IgE production as long as allergen-specific B and T cells are present in the lymph nodes. How inhaled or ingested allergen induces a response specifically in the skin is controversial. Ingested food allergen may reach the circulation in a dose that may be sufficiently high to enter the skin and be picked up by LC. However, whether, for example, house-dust mite allergen, which possesses enzymatic activity, reaches the skin in its un-processed form, or is processed by APC and triggers the response in the skin *via* an indirect route, is still unclear.

Aeroallergen can penetrate the skin directly from the outside, as it occurs in the APT (1). IgE+ LC in the skin can pick up, process and present the allergen in the skin *via* the IgE molecules, bound by the FcεRI and II (2). The allergen can then be presented to allergen-specific memory T cells present in the skin (3). These memory T cells are resident cells from previously resolved AD lesions. The course of events during the very first induction of AD lesions is not completely understood yet. The first AD lesions may derive from degranulation of mast cells, initiating an inflammatory response, by which T cells are attracted as well. When these primary lesions resolve, memory cells may stay in the skin.

Freshly isolated LC from normal skin, and perhaps also from non-lesional AD skin, do not express B7, in contrast to LC in lesional AD skin [94, 95]. Therefore,

Figure 2
Model of the role of T cells in the pathogenesis of atopic dermatitis.
The numbers in the figure correspond to the numbers in the text between brackets. The process of allergen presentation, T cell activation, T cell migration, and factors enhancing these processes are shown. Solid arrows indicate production of factors. Line-dot-line arrows indicate migration. Dotted arrows refer to induction of production or migration.
Abbreviations: ag-spec Th2 cell: antigen-specific T helper 2 cell; bystand Th1 cell: bystander T helper 1 cell; CLA: cutaneous lymphocyte-associated antigen; eo: eosinophil; LC: Langerhans cell; mφ: macrophage; TCR: T cell receptor.

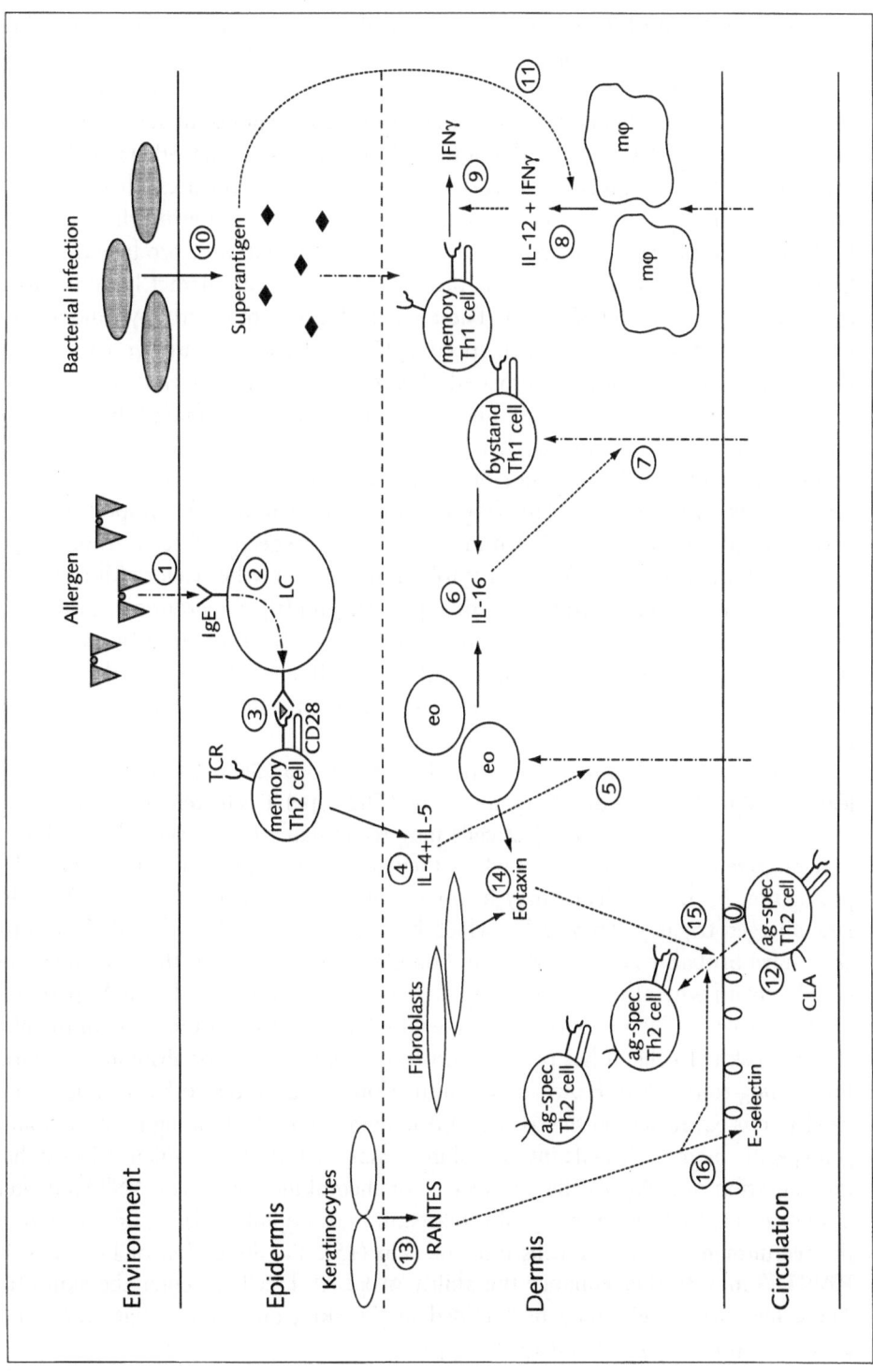

primary activation of T cells by these LC will not be complete, and may induce so-called clonal anergy. Consecutive activation of these T cells may result in the absence of IL-2 and IFNγ production, whereas IL-4 and IL-5 secretion can still occur [96, 97] (4). This may explain the initial Th2 response in acute AD lesions and in the early APT. The production of IL-4 attracts among others eosinophils (5). Although eosinophils are hardly detectable in the infiltrates, their products are present, indicating that eosinophils die rapidly after entering the skin [98]. Early in the response, before dying, the eosinophils may in turn produce IL-16 (6). Later, T cells can contribute to this production. IL-16 can attract large numbers of CD4+ T cells into the skin (7). These CD4+ T cells are mostly bystander cells, which are not necessarily allergen-specific. These cells can be further activated by the action of the previously activated allergen-specific T cells that were already present. The infiltrating eosinophils, and, more probably, macrophages in the infiltrate, may also produce IL-12 (8). The activation of large numbers of bystander T cells in the presence of IL-12 may contribute to the shift towards a Th0/Th1 type response (9). This response follows the initial Th2 response, which was initiated by a much smaller number of allergen-specific T cells. Another factor contributing to the differential cytokine profiles may be bacterial infection of the skin, leading to the production of superantigen (10). Superantigen can down-regulate the production of IL-4, and can induce IFNγ production by monocytes and macrophages (11). *In vivo*, this could also influence the shift of the cytokine profile towards a Th1 response, and contribute to the chronification of AD lesions.

When an allergic skin response resolves, the bystander (Th1-type) T cells will leave the skin. Some of the allergen-specific (Th2-type) T cells remain in the skin as memory cells. Others of the Th2 cells may recirculate. These recirculating T cells may migrate to the lymph nodes where they are involved in the maintenance of IgE production by B cells. They may also recirculate through the body, searching for new allergen triggers. These cells, which have been triggered by the allergen in the skin, have probably acquired the skin homing molecule CLA at that location. This could explain why allergen-specificity in AD patients resides in the CLA+ peripheral blood T cells. The phenotype of these CLA+ T cells also correlates with the phenotype of skin T cells. It has been demonstrated that CLA is not absolutely required for homing to the skin. However, when the skin is triggered at other locations, and E-selectin is expressed on the tissue, CLA may facilitate the homing process of antigen-specific memory T cells into the skin at that location (12), which is faster than the migration of naïve T cells. Production of chemokines, such as RANTES by keratinocytes (13) and eotaxin by fibroblasts and eosinophils (14), can contribute to the recruitment of these T cells into the skin (15). The induction of E-selectin by RANTES may further enhance the ability of CLA+ T cells to enter the skin (16). These memory T cells may be retained in the skin, and may initiate the whole process again upon a new allergen trigger.

In conclusion, the model proposed above implies that, after allergen presentation by LC in the skin of patients with AD, the central role in the initiation, aggravation and continuation of the response towards the allergen is not attributed to mast cells or eosinophils, but to skin-residing T cells.

References

1   Street NE, Mosmann TR (1991) Functional diversity of T lymphocytes due to secretion of different cytokine patterns. *FASEB J* 5: 171–177
2   Mosmann TR, Sad S (1996) The expanding universe of T cell subsets: Th1, Th2 and more. *Immunol Today* 17: 138–146
3   Leung DYM, Bhan AK, Schneeberger EE, Geha RS (1983) Characterization of the mononuclear cell infiltrate in atopic dermatitis using monoclonal antibodies. *J Allergy Clin Immunol* 71: 47–56
4   Langeveld-Wildschut EG, Van Marion AMW, Thepen T, Mudde GC, Bruijnzeel PLB, Bruijnzeel-Koomen CAFM (1995) Evaluation of variables influencing the outcome of the atopy patch test. *J Allergy Clin Immunol* 96: 66–73
5   Langeveld-Wildschut EG, Thepen T, Bihari IC, Van Reijsen FC, De Vries JM, Bruijnzeel PLB, Bruijnzeel-Koomen CAFM (1996) Evaluation of the atopy patch test and the cutaneous late phase reaction, as relevant models for the study of allergic inflammation in patients with atopic eczema. *J Allergy Clin Immunol* 98: 1019–1027
6   Thepen T, Langeveld-Wildschut EG, Bihari IC, Van Wichen DF, Van Reijsen FC, Mudde GC, Bruijnzeel-Koomen CAFM (1996) Biphasic response against aeroallergens in atopic eczema, showing a switch from an initial Th2 response into a Th1 response *in situ*. *J Allergy Clin Immunol* 97: 828–837
7   Bruijnzeel-Koomen CAFM, Van der Donk EMM, Bruijnzeel PLB, Capron M, De Gast GC, Mudde GC (1988) Associated expression of CD1 antigen and the Fc receptor for IgE on Langerhans cells from patients with atopic dermatitis. *Clin Exp Immunol* 74: 137–142
8   Bruijnzeel-Koomen CAFM, Van Wichen DF, Toonstra J, Berrens L, Bruijnzeel PLB (1986) The presence of IgE molecules on epidermal Langerhans cells from patients with atopic dermatitis. *Arch Dermatol Res* 278: 199–205
9   Lanzavecchia A (1987) Antigen uptake and accumulation in antigen-specific B cells. *Immunol Rev* 99: 39–51
10 Wang B, Rieger A, Kilgus O, Ochiai K, Maurer D, Födinger D, Kinet J-P, Stingl G (1992) Epidermal Langerhans cells from normal human skin bind monomeric IgE via FcεRI. *J Exp Med* 175: 1353–1365
11 Bieber T, Rieger A, Neuchrist C, Prinz JC, Rieber EP, Boltz-Nitulescu G, Scheiner O, Kraft D, Ring J, Stingl G (1989) Induction of FcεRII/CD23 on human epidermal Langerhans cells by human recombinant interleukin 4 and γ interferon. *J Exp Med* 170: 309–314

12   Klubal R, Osterhoff B, Wang B, Kinet JP, Maurer D, Stingl G (1997) The high-affinity receptor for IgE is the predominant IgE-binding structure in lesional skin of atopic dermatitis patients. *J Invest Dermatol* 108: 336–342

13   Friedmann PS, Tan BB, Musaba E, Strickland I (1995) Pathogenesis and management of atopic dermatitis. *Clin Exp Allergy* 25: 799–806

14   Mudde GC, Van Reijsen FC, Boland GJ, De Gast GC, Bruijnzeel PLB, Bruijnzeel-Koomen CAFM (1990) Allergen presentation by epidermal Langerhans cells from patients with atopic dermatitis is mediated by IgE. *Immunology* 69: 335–341

15   Kripke ML, Munn CG, Jeevan A, Tang JM, Bucana C (1990) Evidence that cutaneous antigen-presenting cells migrate to regional lymph nodes during contact sensitization. *J Immunol* 145: 2833–2838

16   Macatonia SE, Knight SC, Edwards AJ, Griffiths S, Fryer P (1987) Localisation of antigen on lymph node dendritic cells after exposure to the contact sensitizer fluorescein isothiocyanate. Functional and morphological studies. *J Exp Med* 166: 1654–1667

17   Langeveld-Wildschut EG, Bruijnzeel PLB, Mudde GC, Versluis C, Van Ieperen-Van Dijk AG, Bihari IC, Knol EF, Thepen T, Bruijnzeel-Koomen CAFM, Van Reijsen FC (2000) Clinical and immunological variables in skin of patients with atopic eczema and either positive or negative atopy patch test reactions. *J Allergy Clin Immunol; accepted for publication*

18   Bos JD, Zonneveld I, Das PK, Krieg SR, Van der Loos CM, Kapsenberg ML (1987) The skin immune system (SIS): distribution and immunophenotype of lymphocyte subpopulations in normal human skin. *J Invest Dermatol* 88: 569–573

19   Leung DYM (1995) Atopic dermatitis: the skin as a window into the pathogenesis of chronic allergic diseases. *J Allergy Clin Immunol* 96: 302–318

20   Van Reijsen FC, Bruijnzeel-Koomen CAFM, Kalthoff FS, Maggi E, Romagnani S, Westland JKT, Mudde GC (1992) Skin-derived aeroallergen-specific T cell clones of Th2 phenotype in patients with atopic dermatitis. *J Allergy Clin Immunol* 90: 184–193

21   Kägi MK, Wüthrich B, Montano E, Barandun J, Blaser K, Walker C (1994) Differential cytokine profiles in peripheral blood lymphocyte supernatants and skin biopsies from patients with different forms of atopic dermatitis, psoriasis and normal individuals. *Int Arch Allergy Immunol* 10: 332–340

22   Laberge S, Ghaffar O, Boguniewicz, Center DM, Leung DY, Hamid Q (1998) Association of increased CD4$^+$ T cell infiltration with increased IL-16 gene expression in atopic dermatitis. *J Allergy Clin Immunol* 102: 645–650

23   Center DM, Kornfeld H, Cruikshank WW (1997) Interleukin-16. *Int J Biochem Cell Biol* 29: 1231–1234

24   Lim KG, Wan HC, Bozza PT, Resnick MB, Wong DT, Cruikshank WW, Kornfeld H, Center DM, Weller PF (1996) Human eosinophils elaborate the lymphocyte chemoattractants IL-16 (lymphocyte chemoattractant factor) and RANTES. *J Immunol* 156: 2566–2570

25   Cruikshank WW, Lim K, Theodore AC, Cook J, Fine G, Weller PF, Center DM (1996)

IL-16 inhibition of CD3-dependent lymphocyte activation and proliferation. *J Immunol* 157: 5240–5248

26 Zachary CB, Poulter LW, MacDonald DM (1985) *In situ* quantification of T lymphocyte subsets in atopic dermatitis: the relevance of antigen-presenting cells. *Br J Dermatol* 112: 149–156

27 De Vries IJM, Langeveld-Wildschut EG, Van Reijsen FC, Bihari IC, Bruijnzeel-Koomen CAFM, Thepen T (1997) Non-specific T cell homing during inflammation in atopic dermatitis: expression of cutaneous lymphocyte-associated antigen and integrin αEβ7 on skin-infiltrating T cells. *J Allergy Clin Immunol* 100: 694–701

28 Davis AL, McKenzie JL, Hart DNJ (1988) HLA-DR-positive leucocyte subpopulations in human skin include dendritic cells, macrophages, and CD7-negative T cells. *Immunology* 65: 573–581

29 Silleves Smitt JH, Bos JD, Hulsebosch HJ, Krieg SR (1986) *In situ* immunophenotyping of antigen presenting cells and T cell subsets in atopic dermatitis. *Clin Exp Dermatol* 11: 159–168

30 Wakita H, Sakamoto T, Tokura Y, Takigawa M (1994) E-selectin and vascular cell adhesion molecule-1 as critical adhesion molecules for infiltration of T lymphocytes and eosinophils in atopic dermatitis. *J Cutan Pathol* 21: 33–39

31 Piletta PA, Wirth S, Hommel L, Saurat JH, Hauser C (1996) Circulating skin-homing T cells in atopic dermatitis. Selective up-regulation of HLA-DR, interleukin-2R, and CD30 and decrease after combined UV-A and UV-B phototherapy. *Arch Dermatol* 132: 1171–1176

32 Parada NA, Center DM, Kornfeld H, Rodriguez WL, Cook J, Vallen M, Cruikshank WW (1998) Synergistic activation of CD4+ T cells by IL-16 and IL-2. *J Immunol* 160: 2115–2120

33 Romagnani S (1994) Lymphokine production by human T cells in disease states. *Annu Rev Immunol* 12: 227–257

34 Seder RA, Paul WE, Davis MM, Fazekas de St. Groth B (1992) The presence of Interleukin 4 during *in vitro* priming determines the lymphokine-producing potential of CD4+ T cells from T cell receptor transgenic mice. *J Exp Med* 176: 1091–1098

35 Hsieh C-S, Macatonia SE, Tripp CS, Wolf SF, O'Garra A, Murphy KM (1993) Development of Th1 CD4+ T cells through IL-12 produced by Listeria-induced macrophages. *Science* 260: 547–549

36 Punnonen J, Yssel H, De Vries JE (1997) The relative contribution of IL-4 and IL-13 to human IgE synthesis induced by activated CD4+ or CD8+ T cells. *J Allergy Clin Immunol* 100: 792–801

37 André F, Pène J, André C (1996) Interleukin-4 and interferon-gamma production by peripheral blood mononuclear cells from food-allergic patients. *Allergy* 51: 350–355

38 Koning H, Neijens HJ, Baert MRM, Oranje AP, Savelkoul HFJ (1997) T cell subsets and cytokines in allergic and non-allergic children. I. Analysis of IL-4, IFN-γ and IL-13 mRNA expression and protein production. *Cytokine* 9: 416–426

39 Hamid Q, Boguniewicz M, Leung DYM (1994) Differential in situ cytokine gene expression in acute versus chronic atopic dermatitis. *J Clin Invest* 94: 870–876

40 Hamid Q, Naseer T, Minshall EM, Song YL, Boguniewicz M, Leung DYM (1996) *In vivo* expression of IL-12 and IL-13 in atopic dermatitis. *J Allergy Clin Immunol* 98: 225–231

41 Grewe M, Walther S, Gyufko K, Czech W, Schöpf E, Krutmann J (1995) Analysis of the cytokine pattern expressed *in situ* in inhalant allergen patch test reactions of atopic dermatitis patients. *J Invest Dermatol* 105: 407–410

42 Werfel T, Morita A, Grewe M, Renz H, Wahn U, Krutmann J, Kapp A (1996) Allergen specificity of skin-infiltrating T cells is not restricted to a type-2 cytokine pattern in chronic skin lesions of atopic dermatitis. *J Invest Dermatol* 107: 871–876

43 Grewe M, Gyufko K, Schöpf E, Krutmann J (1994) Lesional expression of interferon-γ in atopic eczema. *Lancet* 343: 25–26

44 Grewe M, Bruijnzeel-Koomen CAFM, Schöpf E, Thepen T, Langeveld-Wildschut AG, Ruzicka T, Krutmann J (1998) A role for Th1 and Th2 cells in the immunopathogenesis of atopic dermatitis. *Immunol Today* 19: 359–361

45 Grewe M, Czech W, Morita A, Werfel T, Klammer M, Kapp A, Ruzicka T, Schöpf E, Krutmann J (1998) Human eosinophils produce biologically active IL-12: Implications for control of T cell responses. *J Immunol* 161: 415–420

46 Vezzio N, Sarfati M, Yang LP, Demeure CE, Delespesse G (1996) Human Th2-like cell clones induce IL-12 production by dendritic cells and may express several cytokine profiles. Int Immunol 8: 1963–1970

47 Bohle B, Schwila H, Hu HZ, Friedl-Hajek R, Sowka S, Ferreira F, Breiteneder H, Bruijnzeel-Koomen CA, De Weger RA, Mudde GC, Ebner C, Van Reijsen FC (1998) Long-lived Th2 clones specific for seasonal and perennial allergens can be detected in blood and skin by their TCR-hypervariable regions. *J Immunol* 160: 2022–2027

48 Bouchonnet F, Lecossier D, Bellocq A, Hamy I, Hance AJ (1994) Activation of T cells by previously activated T cells. HLA-unrestricted alternative pathway that modifies their proliferative potential. *J Immunol* 153: 1921–1935

49 Miller RA, Coleman CN, Fawcett HD, Hoppe RT, McDougall IR (1980) Sezary syndrome: a model for migration of T lymphocytes to skin. *N Engl J Med* 303: 89–92

50 Picker LJ, Michie SA, Rott LS, Butcher EC (1990) A unique phenotype of skin-associated lymphocytes in humans. Preferential expression of the HECA-452 epitope by benign and malignant T cells at cutaneous sites. *Am J Pathol* 136: 1053–1068

51 Zollner TM, Nuber V, Duijvestijn AM, Boehncke WH, Kaufmann R (1997) Superantigens but not mitogens are capable of inducing upregulation of E-selectin ligands on human T lymphocytes. *Exp Dermatol* 6: 161–166

52 Berg EL, Robinson MK, Mansson O, Butcher EC, Magnani JL (1991) A carbohydrate domain common to both sialyl Le a and sialyl Le x is recognized by the endothelial cell leukocyte adhesion molecule ELAM-1. *J Biol Chem* 266: 14869–14872

53 Fuhlbrigge RC, Kieffer JD, Armerding D, Kupper TS (1997) Cutaneous lymphocyte

antigen is a specialized form of PSGL-1 expressed on skin-homing T cells. *Nature* 389: 978–981

54   Kansas GS (1996) Selectins and their ligands: current concepts and controversies. *Blood* 88: 3259–3287

55   Rossiter H, Van Reijsen F, Mudde GC, Kalthoff F, Bruijnzeel-Koomen CA, Picker LJ, Kupper TS (1994) Skin disease-related T cells bind to endothelial selectins: expression of cutaneous lymphocyte antigen (CLA) predicts E-selectin but not P-selectin binding. *Eur J Immunol* 24: 205–210

56   Hirata N, Kohrogi H, Iwagoe H, Goto E, Hamamoto J, Fujii K, Yamaguchi T, Kawano O, Ando M (1998) Allergen exposure induces the expression of endothelial adhesion molecules in passively sensitized human bronchus: time course and the role of cytokines. *Am J Respir Cell Mol Biol* 18: 12–20

57   Adams DH, Hubscher SG, Fischer NC, Williams A, Robinson M (1996) Expression of E-selectin and E-selectin ligands in human liver inflammation. Hepatology 24: 533–538

58   Brockmeyer C, Ulbrecht M, Schendel DJ, Weiss EH, Hillebrand G, Burkhardt K, Land W, Gokel MJ, Riethmuller G, Feucht HE (1993) Distribution of cell adhesion molecules (ICAM-1, VCAM-1, ELAM-1) in renal tissue during allograft rejection. *Transplantation* 55: 610–615

59   Borowitz MJ, Weidner A, Olsen EA, Picker LJ (1993) Abnormalities of circulating T cell subpopulations in patients with cutaneous T cell lymphoma: cutaneous lymphocyte-associated antigen expression on T cells correlates with extent of disease. *Leukemia* 7: 859–863

60   Jones SM, Dixey J, Hall ND, McHugh NJ (1997) Expression of the cutaneous lymphocyte antigen and its counter-receptor E-selectin in the skin and joints of patients with psoriatic arthritis. *Br J Rheumatol* 36:748–757

61   Gonzalez FJ, Carvajal MJ, Leiva L, Juarez C, Blanca M, Santamaria LF (1997) Expression of the cutaneous lymphocyte-associated antigen in circulating T cells in drug-allergic reactions. *Int Arch Allergy Immunol* 113: 345–347

62   Abernathy-Carver KJ, Sampson HA, Picker LJ, Leung DY (1995) Milk-induced eczema is associated with the expansion of T cells expressing cutaneous lymphocyte antigen. *J Clin Invest* 95: 913–918

63   Santamaria Babi LF, Perez Soler MT, Hauser C, Blaser K (1995) Skin-homing T cells in human cutaneous allergic inflammation. *Immunol Res* 14: 317–324

64   Santamaria LF, Perez Soler MT, Hauser C, Blaser K (1995) Allergen specificity and endothelial transmigration of T cells in allergic contact dermatitis and atopic dermatitis are associated with the cutaneous lymphocyte antigen. *Int Arch Allergy Immunol* 107: 359–362

65   Kuijpers TW, Etzioni A, Pollack S, Pals ST (1997) Antigen-specific immune responsiveness and lymphocyte recruitment in leukocyte adhesion deficiency type II. *Int Immunol* 9: 607–613

66   Strunk D, Egger C, Leitner G, Hanau D, Stingl G (1997) A skin homing molecule defines the Langerhans cell progenitor in human peripheral blood. *J Exp Med* 185: 1131–1136

67 Koszik F, Strunk D, Simonitsch I, Picker LJ, Stingl G, Payer E (1994) Expression of monoclonal antibody HECA-452-defined E-selectin ligands on Langerhans cells in normal and diseased skin. *J Invest Dermatol* 102: 773–780

68 Sallusto F, Lanzavecchia A, Mackay CR (1998) Chemokines and chemokine receptors in T cell priming and Th1/Th2-mediated responses. *Immunol Today* 19: 568–574

69 Sallusto F, Mackay CR, Lanzavecchia A (1997) Selective expression of the eotaxin receptor CCR3 by human T helper 2 cells. *Science* 277: 2005–2007

70 Gerber BO, Zanni MP, Uguccioni M, Loetscher M, Mackay CR, Pichler WJ, Yawalkar N, Baggiolini M, Moser B (1997) Functional expression of the eotaxin receptor CCR3 in T lymphocytes co-localizing with eosinophils. *Curr Biol* 7: 836–843

71 Crump MP, Rajarathnam K, Kim KS, Clark-Lewis I, Sykes BD (1998) Solution structure of eotaxin, a chemokine that selectively recruits eosinophils in allergic inflammation. *J Biol Chem* 273: 22471–22479

72 Teixeira MM, Wells TN, Lukacs NW, Proudfoot AE, Kunkel SL, Williams TJ, Hellewell PG (1997) Chemokine-induced eosinophil recruitment. Evidence of a role for endogenous eotaxin in an *in vivo* allergy model in mouse skin. *J Clin Invest* 100: 1657–1666

73 Mochizuki M, Bartels J, Mallet AI, Christophers E, Schröder JM (1998) IL-4 induces eotaxin: a possible mechanism of selective eosinophil recruitment in helminth infection and atopy. *J Immunol* 160: 60–68

74 Yamada H, Matsukura M, Yudate T, Chihara J, Stingl G, Tezuka T (1997) Enhanced production of RANTES, an eosinophil chemoattractant factor, by cytokine-stimulated epidermal keratinocytes. *Int Arch Allergy Immunol* 114 (1): 28–32

75 Taguchi M, Sampath D, Koga T, Castro M, Look DC, Nakajima S, Holtzman MJ (1998) Patterns for RANTES secretion and intercellular adhesion molecule 1 expression mediate transepithelial T cell traffic based on ion analyses *in vitro* and *in vivo*. *J Exp Med* 187: 1927–1940

76 Gerber BO, Zanni MP, Uguccioni M, Loetscher M, Mackay CR, Pichler WJ, Yawalkar N, Baggliolini M, Moser B (1997) Functional expression of the eotaxin receptor CCR3 in T lymphocytes co-localizing with eosinophils. *Curr Biol* 7: 836–843

77 Jinquan T, Quan S, Feili G, Larsen CG, Thestrup-Pedersen K (1999) Eotaxin activates T cells to chemotaxis and adhesion only if induced to express CCR3 by IL-2 together with IL-4. *J Immunol* 162: 4285–4292

78 Beck LA, Dalke S, Leiferman KM, Bickel CA, Hamilton R, Rosen H, Bochner BS, Schleimer RP (1997) Cutaneous injection of RANTES causes eosinophil recruitment: comparison of nonallergic and allergic human subjects. *J Immunol* 159: 2962–2972

79 Leyden JE, Marpies RR, Klingman AM (1974) Staphylococcus aureus skin colonization in atopic dermatitis. *Br J Dermatol* 90: 525–530

80 Lever R, Hadley K, Downey D, Mackie R (1988) Staphylococcal colonization in atopic dermatitis and the effect of topical mupirocin therapy. *Br J Dermatol* 119: 189–198

81 Herman A, Kappler JW, Marrack P, Pullen AM (1991) Superantigens: mechanism of T cell stimulation and role in immune responses. *Annu Rev Immunol* 9: 745–772

82 Marrack P, Kappler J (1990) The staphylococcal enterotoxins and their relatives. *Science* 248: 705–711

83 Strange P, Skov L, Lisby S, Nielsen PL, Baadsgaard O (1996) Staphylococcal enterotoxin B applied on intact normal and intact atopic skin induces dermatitis. *Arch Dermatol* 132: 27–33

84 Torres MJ, Gonzalez FJ, Corzo JL, Giron MD, Carvajal MJ, Garcia V, Pinedo A, Martinez-Valverde A, Blanca M, Santamaria LF (1998) Circulating CLA⁺ lymphocytes from children with atopic dermatitis contain an increased percentage of cells bearing staphylococcal-related T cell receptor variable segments. *Clin Exp Allergy* 28: 1264–1272

85 Strickland I, Hauk PJ, Trumble AE, Picker LJ, Leung DYM (1999) Evidence for superantigen involvement in skin homing of T cells in atopic dermatitis. *J Invest Dermatol* 112: 249–253

86 Neuber K, Steinrucke K, Ring J (1995) Staphylococcal enterotoxin B affects *in vitro* IgE synthesis, interferon-gamma, interleukin-4 and interleukin-5 production in atopic eczema. *Int Arch Allergy Immunol* 107: 179–182

87 Campbell DE, Fryga AS, Bol S, Kemp AS (1999) Intracellular interferon-gamma (IFN-γ) production in normal children and children with atopic dermatitis. *Clin Exp Immunol* 115: 377–382

88 Bunikowski R, Mielke M, Skarabis H, Herz U, Bergmann RL, Wahn U, Renz H (1999) Prevalence and role of serum IgE antibodies to the Staphylococcus aureus-derived superantigens SEA and SEB in children with atopic dermatitis. *J Allergy Clin Immunol* 103: 119–124

89 Hofer MF, Harbeck RJ, Schlievert PM, Leung DY (1999) Staphylococcal toxins augment specific IgE responses by atopic patients exposed to allergen. *J Invest Dermatol* 112: 171–176

90 König B, Neuber K, König W (1995) Responsiveness of peripheral blood mononuclear cells from normal and atopic donors to microbial superantigens. *Int Arch Allergy Immunol* 106: 124–133

91 Herz U, Schnoy N, Borelli S, Weigl L, Käsbohrer U, Daser A, Wahn U, Köttgen F, Renz H (1998) A human-SCID mouse model for allergic immune responses: bacterial superantigen enhances skin inflammation and suppresses IgE production. *J Invest Dermatol* 110: 224–231

92 Lester MR, Hofer MF, Renz H, Trumble AE, Gelfand EW, Leung DY (1995) Modulatory effects of staphylococcal superantigen TSST-1 on IgE synthesis in atopic dermatitis. *Clin Immunol Immunopathol* 77: 332–338

93 Mandel TE, Phipps RP. Abbott A, Tew JG (1980) The follicular dendritic cell: long term antigen retention during immunity. *Immunol Rev* 53: 29–57

94 Symington FW, Brady W, Linsley PS (1993) Expression and function of B7 on human epidermal Langerhans cells. *J Immunol* 150: 1286–1295

95 Ohki O, Yokozeki H, Katayama I, Umeda T, Azuma M, Okumura K, Nishioka K (1997) Functional CD86 (B7-2/B70) is predominantly expressed on Langerhans cells in atopic dermatitis. *Br J Dermatol* 136: 838–845

96 Van Reijsen FC, Wijburg OLC, Gebhardt M, Van Ieperen-Van Dijk AG, Betz S, Poellabauer E-M, Thepen T, Bruijnzeel-Koomen CAFM, Mudde GC (1994) Different growth factor requirements for human Th2 cells may reflect *in vivo* induced anergy. *Clin Exp Immunol* 98: 151–157

97 Gajewski TF, Lancki DW, Stack R, Fitch FW (1994) "Anergy" of Th0 helper T lymphocytes induces downregulation of Th1 characteristics and a transition to a Th2-like phenotype. *J Exp Med* 179: 481–491

98 Bruijnzeel-Koomen CAFM, Van Wichen DF, Spry CJF, Venge P, Bruijnzeel PLB (1988) Active participation of eosinophils in patch test reactions to inhalant allergens in patients with atopic dermatitis. *Br J Dermatol* 118: 229–238

Immunological aspects of allergic inflammation: eosinophils

Jörn Elsner and Alexander Kapp

Hannover Medical University, Department of Dermatology and Allergology, Ricklinger Str. 5, 30449 Hannover, Germany

Introduction

Increased numbers of eosinophils in the peripheral blood and inflammatory tissue are characteristic features of allergic diseases such as allergic asthma, rhinoconjunctivitis and atopic dermatitis. In addition, eosinophils are believed to be of major importance in other inflammatory diseases such as connective tissue diseases of unknown origin such as hypereosinophilic syndromes and bullous dermatoses. Tissue damage and propagation of inflammation is thought to be mediated by the interaction between Th2-like T cells, antigen-presenting cells and eosinophils. In this process, eosinophils are activated by several inflammatory mediators leading to invasion of eosinophils at the site of inflammation and to tissue damaging by the release of reactive oxygen species and toxic granule proteins.

Cytokines, chemokines and inflammatory mediators, such as GM-CSF, IL-5, eotaxin, PAF C5a and C3a are responsible for the activation of human eosinophils. Under the high shear forces present in the blood flow, eosinophils become first tethered and then roll along the vessel surface. When local signals such as cytokines or chemokines are released in their vicinity, cells arrest, develop firm adhesion and then migrate across the endothelium. Thereafter, cells orientate their locomotion along a gradient of a chemotactic factor by extending lamellipodia in the direction of the higher concentration of the chemotactic factor. Migration requires the coordination of a cycle of cytoskeletal proteins such as actin entailing the formation of adhesive contacts at the leading edge of the cell, breaking adhesive contacts and cytoskeleton-dependent retraction at the trailing edge. In a previous study, we could demonstrate that intracellular Ca^{2+} fluxes in eosinophils following stimulation with C5a, PAF and RANTES represent an important step in the activation leading to directed migratory response and actin polymerization.

At the site of inflammation eosinophils release toxic proteins such as eosinophilic cation protein (ECP), major basic protein (MBP), eosinophil-derived neurotoxin (EDN) and reactive oxygen species leading to tissue damage of the host. Reactive oxygen species are generated by the NADPH oxidase that can be activated by a number of different soluble and particulate agents. Intracellular Ca^{2+} seems to play a central role in the modulation of the respiratory burst in eosinophils. The activation results in a reduction of molecular oxygen to the potentially toxic oxygen species superoxide anion ($\cdot O_2^-$) or hydrogen peroxide (H_2O_2), with NADPH serving as the electron donor. The increase in oxygen consumption is termed the respiratory burst and acts as a major contribution to tissue damage and propagation of the inflammatory response by acting as a competence signal in T lymphocytes, inducing early gene expression, particularly IL-2, as well as cell proliferation.

Until now, there exists no specific therapy to hinder the eosinophil response in allergic and non-allergic diseases. The treatment of patients with steroids, dapson and H_1-receptor antagonists has been reported in eosinophilic diseases, however, these compounds have also been effective on other cells. Therefore, agents that would be able to inhibit or antagonize mediator-induced eosinophil activation seem to be interested as new therapeutical strategy. In this chapter, we will focus on the functional properties and the modulation of human eosinophils in inflammation. We will then discuss whether modulation of eosinophil effector functions might be successful as a future therapeutic strategy of diseases that are accompanied with activated eosinophils.

Eosinophils and apoptosis

Hypereosinophilic diseases such as allergic asthma and atopic dermatitis are accompanied by elevated number of eosinophils in the peripheral blood or in the inflamed tissue, possibly due to an enhanced eosinophil survival. *In vitro* studies have demonstrated that cytokines such as IL-3, IL-5, and GM-CSF result in a prolonged eosinophil survival by abrogating apoptosis or programmed cell death [1–3]. Apoptosis is an ordered process of fundamental biologic importance characterized by cell morphological changes, DNA fragmentation and loss of nucleoli. Aged apoptotic eosinophils were recognized and ingested as intact cells by autologous macrophages. The mechanism of abrogating apoptosis by IL-3, IL-5, and GM-CSF has recently been demonstrated [4]. Anti-apoptotic effect of these cytokines seems to be associated by interactions of their β-receptor with tyrosine kinases particularly by sequential activation of Lyn and Syk tyrosine kinases [5]. Recent studies have demonstrated that cross-linking of CD95 (Fas antigen) or CD69 induce apoptosis of human eosinophils [1, 2, 6–9]. Moreover, the steroid dexamethason has been shown to induce apoptosis of human eosinophils while inhibiting neutrophil apoptosis [1, 10]. Therefore, apoptosis of human eosinophils may contribute to the resolution of inflammatory reactions in which eosinophil accumulation is a prominent feature. In this context, in a previous study it has been shown that eosinophil apoptosis is markedly delayed in the so-called atopic diseases irrespective of allergen sensitization and suggests that this effect is mediated by the autocrine production of growth factors by eosinophils [4].

Surface molecules expressed by human eosinophils

General surface receptors

Many of the interactions between inflammatory cells and their environment are mediated by surface molecules which bind both intracellular and extracellular ligands and thus transfer signals across the membrane (Fig. 1). These signals may result in changes in cytoskeletal morphology, gene expression, cellular differentiation and secretion. Eosinophils share some surface molecules with other granulocytes such as adhesion molecule receptors [11–17], Fc receptors [18–20], cytokine receptors, chemokine receptors [21–25] and complement receptors [26]. Surface markers have been used to distinguish neutrophils from eosinophils and might indicate different functional properties of these cells.

CD69 is expressed on activated eosinophils

An interesting molecule on the surface of eosinophils is the CD69 antigen, an activation antigen that is also found on activated neutrophils, lymphocytes and NK cells [7, 27, 28]. CD69 is not constitutively expressed on freshly isolated eosinophils, but culture of eosinophils with GM-CSF, IL-3 or IL-5 induces CD69 expression (Fig. 1). Moreover, eosinophils in bronchoalveolar lavage fluid from patients with asthma and pulmonary eosinophilia also express CD69 [27, 28]. More recently, it has been demonstrated that ligation of CD69 induces apoptosis and cell death in human eosinophils cultured with GM-CSF and IL-13 [7]. Therefore, this antigen might play an important role for eosinophil removal *in vivo* [6].

Fc-receptors on eosinophils

Controversial results have been described about Fcε-receptor expression on human eosinophils [18, 29–33]. The constitutive expression of the low affinity IgE receptor (Fcε-RII/CD23) has been demonstrated on mRNA level on human eosinophils, however flow cytometric analysis shows only few and heterogeneous expression of Fcε-RII/CD23 receptors [18,3 0]. In addition, of most interest becomes the observation the eosinophils also express the high-affinity IgE receptor (Fcε-RI) that is involved in degranulation and antigen presentation of mast cells and Langerhans cells [33, 34]. However, studies in which flow cytometry or immunostaining was used did not reveal surface expression of Fcε-RI protein on human eosinophils [35]. Only mRNA of Fcε-RI in eosinophils was detected in several studies [35]. Therefore, it is doubtful whether eosinophils express Fcε-RI because mRNA detection of this antigen could only be due to contamination of other cell types such as

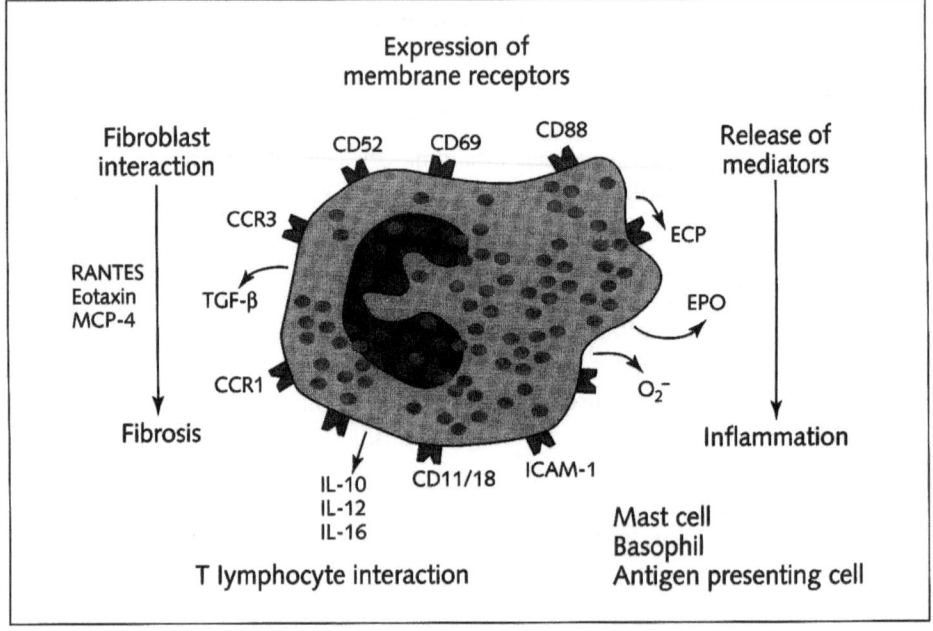

Figure 1

Eosinophil interaction.

Eosinophils are able to interact with different cell types by surface molecules, cytokines and chemokines.

basophils (when using blood) or Langherhans cells and mast cells (when using tissue).

CD52 is a new eosinophil differentiation antigen

Therapeutic strategies to prevent selectively the destructive power of neutrophils and eosinophils have been hampered by the lack of surface molecules that are involved in only eosinophil or neutrophil activation. Therefore it seems to be more important to find surface molecules that are expressed on a single cell type only. Interesting molecules for that view are the recently published surface antigen CD40 ligand (CD40L) and CD40 that is originally identified on the surface of B lymphocytes [36, 37]. The importance of CD40-CD40 ligand interactions has been recently demonstrated in allergic diseases because it is thought to be involved in the switching of B lymphocytes to an IgE phenotype. More recently, CD40 surface and mRNA expression has been demonstrated in human eosinophils. Cross-linking of this antigen results in enhanced eosinophil survival and the release of GM-CSF [37]. Another molecule that becomes of interest is the adhesion molecule CD49d (VLA-4) that is expressed on the surface of eosinophils [13, 14, 17]. It has been demonstrated that CD49d/VCAM-1 interaction plays a predominant role in controlling antigen-induced eosinophil recruitment into the tissue. However, treatment with anti-CD49d mAb alone was not able to prevent eosinophil invasion into the inflammatory tissue [13, 14, 17]. Another surface molecule that has been reported to distinguish eosinophils from neutrophils is the surface antigen CD9 [38]. However, flow cytometric analysis clearly demonstrates that this antigen is also expressed on neutrophils (own unpublished results).

Recently, we have been searching for a surface molecule expressed only on eosinophils but not on neutrophils (Fig. 1). We could demonstrate that CD52, a GPI-anchored molecule is expressed on the surface and produced on mRNA level by human eosinophils, but not by neutrophils [26]. The CD52 antigen first identified on T lymphocytes and macrophages is an excellent good target for complement-mediated attack. For this reason anti-CD52 mAbs have been widely used for removal of T cells from donor bone marrow to prevent graft-*versus*-host disease in humans [39–41]. Anti-CD52 mAbs are therefore suitable to distinguish eosinophils from neutrophils as a simple marker molecule in flow cytometry but also to enrich neutrophils from hypereosinophilic patients. Furthermore, the importance of the CD52 molecule on human eosinophils has been shown by its function as an inhibitory signal on the respiratory burst following cross-linking [26]. Therefore, this antibody seems to be an interesting tool in the therapy of diseases that are accompanied by blood and tissue eosinophilia (Fig. 3).

The complement system and eosinophils

Activation of eosinophils by C3a

The complement system is a major element of the humoral defense reaction. One group of highly active mediators that are generated by complement activation are the anaphylatoxins C3a, C4a, and C5a [42]. During an inflammatory process, local production of complement-derived mediators, e.g., due to the release of mast cell derived tryptase, results in increased vascular permeability, leukocyte adherence and directed migration of leukocytes into the site of inflammation [43]. The complement-derived anaphylatoxins C3a and C5a bind to specific cellular receptors and thereby trigger subsequent cellular responses [44, 45]. A well known function of C3a is the induction of smooth muscle contraction, the release of histamine from mast cells and the activation of guinea-pig platelets [42]. The role of C3a in the modulation of leukocyte functions, in contrast to its chemical analogues C5a, is poorly understood and controversially discussed. Earlier studies reported that C3a induced granule release, chemotaxis and aggregation of neutrophils [46, 47]. However, most of the earlier reported effects of C3a have been hampered due to the contamination of C3a preparations with C5a. Previous studies revealed biological effects of purified C3a, besides C5a, in the activation of the respiratory burst in human eosinophils [48, 49]. This activation was coupled specifically to the C3a receptor. Stimulation of eosinophils by C3a involves pertussis toxin-sensitive Gi-proteins and leads to a transient rise in $[Ca^{2+}]_i$. Therefore, C3a plays an important role as a potent mediator for the microbicidal and tissue destructive power of eosinophils and may take part in the pathophysiology of diseases with complement activation and activated eosinophils, such as inflammatory skin diseases [48, 49].

C5a receptor is expressed by human eosinophils

Apparently, C5a seems to be the most potent pro-inflammatory mediator derived from the complement system [50, 51]. Besides inducing the release of granule enzymes and reactive oxygens species, C5a also acts as a chemotaxin for neutrophils and eosinophils [48, 49, 52]. Since C3a is not chemotactic for eosinophils (own unpublished results), C5a is thought to responsible for the infiltration of eosinophils in the tissue. Moreover, C5a represents a major metabolite activator for eosinophils inducing the release of toxic granule proteins and reactive oxygen species that cause damage to the host tissue [53, 54]. Recently, different antibodies against the C5a receptor were produced [50, 55, 56]. A polyclonal anti-C5a receptor antibody was able to prevent zymosan-activated serum (ZAS)-induced neutrophil chemotaxis and the monoclonal anti-C5a receptor mAb S5/1 specifically inhibited C5a-induced intracellular calcium transients in human neutrophils [55, 56]. In contrast to the well characterized neutrophil C5a receptor, which was recently identified as a 350-residue G protein-coupled receptor, there are only few data on the eosinophil C5a receptor [57]. Scatchard analyses of C5a binding to human eosinophils have implied the existence of high and low affinity binding sites on eosinophils [58].

In a previous study it could be demonstrated that the anti-C5a receptor mAb S5/1 detected homogeneous C5a receptor expression on the surface of human eosinophils and completely inhibited several eosinophil effector functions [59]. The human eosinophil C5a receptor is homogeneously expressed not only on normal eosinophils from healthy donors, but also on hypodense and normodense eosinophil subpopulations which were obtained from patients with hypereosinophilia [59]. Based on the inhibitory effect of the S5/1 mAb on C5a-induced eosinophil effector functions, a single receptor type, identified by this mAb, mediates C5a effects in these cells. In addition, the inhibitory effect of the S5/1 mAb on C5a functions may enable a new experimental approach to the treatment of diseases that have been associated with C5a-mediated activation (Fig. 3).

Eosinophils are activated by cytokines and chemokines

Cytokines

Eosinophils are produced in the bone marrow under the stimulation of differentiation factors, such as IL-3, IL-5 and GM-CSF [60–63]. These cytokines are also involved in prolonged eosinophil survival in culture by abrogating apoptosis. Moreover, they are involved in several eosinophil effector functions such as the activation of the respiratory burst [20, 60, 64].

Chemokines activate eosinophils

Besides several mediators such as C5a, C3a, PAF, and 5-oxo-eicosanoids [48, 49, 53, 65, 66], in the last decade, a new family of chemotactic cytokines, now termed chemokines, have become interesting because of their restricted target cell specificity [67–72]. Until now there are four chemokine subfamilies which can distinguished on the basis of the arrangement of the amino acid cysteine in the amino-terminal region: CXC chemokines, CC chemokines, C chemokines and CXXXC chemokines [68, 69]. The chemokine subclasses differ in their biologic activity to stimulate different kinds of effector cells [67, 68]. Chemokines of the CXC chemokine family such as IL-8, ENA-78, GRO-α and GCP-2 activate predominantly neutrophils [67, 68, 73–81]. Chemokines of the CC chemokine family (Tab. 1), such as RANTES, MIP-1α, MCP-2, MCP-3, MCP-4, eotaxin and eotaxin-2 activate predominantly eosinophils [64, 67, 68, 82–86]. The role of the recently cloned CC chemokines AMAC-1, leukotactin-1, exodus-1,2,3 on eosinophils has not been investigated so far [87–91].

Recent studies demonstrated that CC chemokines are not only potent activators of eosinophil chemotaxis, but also involved in the activation of the respiratory burst [52, 64, 92–95]. The CC chemokines eotaxin, eotaxin-2, MCP-2, MCP-3, MCP-4 and RANTES represent potent activators of eosinophil effector functions with clearly distinct profiles of activity. These studies underline the importance of eotaxin as a potent activator of the respiratory burst (eotaxin = eotaxin-2 > MCP-3 = MCP-4 > RANTES), actin polymerization and chemotaxis (eotaxin = eotaxin-2 = RANTES > MCP-3 =MCP-4 [52, 94, 95].

A more recent study revealed the importance of CC chemokine profile in the skin of human atopic subjects [96]. Intradermally timothy grass pollen extract allergen challenge results in a significant increase in mRNA[+] cells for MCP-3, which peaked at 6 h and progressively declined at 24 and 48 h. This parallels the kinetics of total (MBP[+] cells) and activated (EG2[+] cells) eosinophil infiltration [96]. The allergen-induced expression of mRNA[+] for RANTES is also observed at 6 h, with a maximum at 24 h and decrease at 48 h. Moreover, the number of mRNA[+] cells for

Table 1 - CC chemokine and CC chemokine receptors (CCR)

Receptor	Ligand	Expression
CCR1	MIP-1α, RANTES, MCP-3, MCP-4	monocytes, eosinophils, T cells
CCR2a/2b (mRNA-splice variants)	MCP-1, MCP-3, MCP-4	monocytes, T cells, basophils, B cells
CCR3	RANTES, eotaxin, eotaxin-2, eotaxin-3, MCP-3, MCP-4	eosinophils, Th2 cells, basophils
CCR4	MIP-1α, RANTES, MCP-1, TARC, Exodus-1	basophils, T cells (Th2)
CCR5	MIP-1α, MIP-1β, RANTES	monocytes, T cells (Th1, Th2)
CCR6	MIP-1α, MIP-1β, RANTES, MCP-3, MCP-1 (?), Exodus-1	monocytes, T cells
CCR7 (EBI1, BLR-2)	Exodus-2, Exodus-3, PARC	activated B cells and T cells
CCR8 (TER1)	I-309, TCA-3, TARC	activated T cells
CCR9 or CCR10 or D6	MIP-1α, MIP-1β, RANTES, MCP-1,2,3,4, eotaxin	placenta, fetal liver, spleen

Abbreviations: MCP, monocyte chemotactic protein; RANTES, regulated upon activation, normal T cell expressed and secreted; MIP, macrophage inflammatory protein-1α; TARC, thymus and activation-regulated chemokine; LARC, liver and activation-regulated chemokine (= Exodus-1); PARC, pulmonary and activation-regulated chemokine; EBI, Epstein Barr-induced; ELC, EBI1-ligand chemokine (= Exodus-3); SLC, secondary lymphoid-tissue chemokine (Exodus-2)

RANTES parallels the kinetics of infiltration of CD3+, CD4+ and CD8+ T cells, whereas the number of CD68+ macrophages was still increasing at 48 h. Therefore, these data suggest that MCP-3 seems to be involved in the regulation of the early eosinophil response to specific allergen, whereas RANTES may be involved in the later accumulation of T cells and macrophages [96]. Moreover, eosinophil recruitment occurred more rapidly in allergic subjects than in non-allergic subjects following intradermal injection of RANTES [97].

Chemokine receptors

According to desensitization experiments, binding studies and mRNA detection, normal eosinophils seem to express two CC chemokine receptors, CCR1 and CCR3 [23, 24, 94, 98–103]. MIP-1α, RANTES and MCP-3 have been identified as ligands for CCR1, and RANTES, MCP-3, MCP-4 and eotaxin for CCR3 (Tab. 1). Recently, desensitization experiments between eotaxin and eotaxin-2 and binding studies with an anti-CCR3 mAb gave evidence that both CC chemokines act through the CCR3 [95, 104, 105]. The usage of monoclonal antibodies against human eotaxin revealed that certain leukocytes, as well as respiratory epithelium, were intensely immunoreactive, and eosinophil infiltration occurred at sites of eotaxin upregulation [100]. Moreover, increased levels of mRNA of human eotaxin were found in normal small bowel and colon and at lower levels and other organs [24, 106]. The importance of CCR3 has recently been demonstrated in a study of patients with allergic asthma: Eotaxin (source: epithelial and endothelial cells) and CCR3 mRNA and protein were significantly elevated in bronchial mucosal biopsies from patients with allergic asthma compared with controls [107]. Moreover, differential expression of chemokine receptors and chemotactic responsiveness of naïve T cells (CXCR4+), majority of memory/activated T cells (CXCR3+), Th0 (CXCR3+), Th1 (CXCR3+ and CCR5+) and Th2 (CCR3+, CCR4+, CCR5+) CD4-positive cells could be demonstrated [108–111].

Eosinophils synthesize and release chemokines and cytokines

Recent studies demonstrated that eosinophils are also able to synthesize cytokines and chemokines, such as IL-6, IL-8, RANTES, and IL-16 [112-117] (Fig. 2). Thus, eosinophils represent a source of cytokines that are chemoattractants for neutrophils, lymphocytes, as well as eosinophils. Therefore, eosinophils could contribute cytokines to enhance the recruitment of additional populations of lymphocytes and eosinophils. Other important eosinophil-derived cytokines are TGFβ, which might be important for the interaction with fibroblasts in wound healing and connective tissue diseases such as scleroderma [118, 119]. IL-10 and IL-12 from eosinophils might be able to suppress and activate T cells [120, 121]. Therefore, eosinophils are not only potent effector cells due to their capacity to secrete toxic proteins and reactive oxygen species, but also responsible for the propagation of the inflammatory response due to the release of chemokines and cytokines (Fig. 1 and 2).

Modulation of eosinophil effector functions

The role of intracellular calcium fluxes in eosinophil effector functions

It is well know that transient elevation of intracellular calcium concentration ($[Ca^{2+}]_i$) plays an important role as a second messenger in many cell types [122–125]. Previously, the role of $[Ca^{2+}]_i$ transients in the activation of human eosinophils was investigated. In one study it could be demonstrated that intracellular Ca^{2+} fluxes in eosinophils are important for the activation of the respiratory

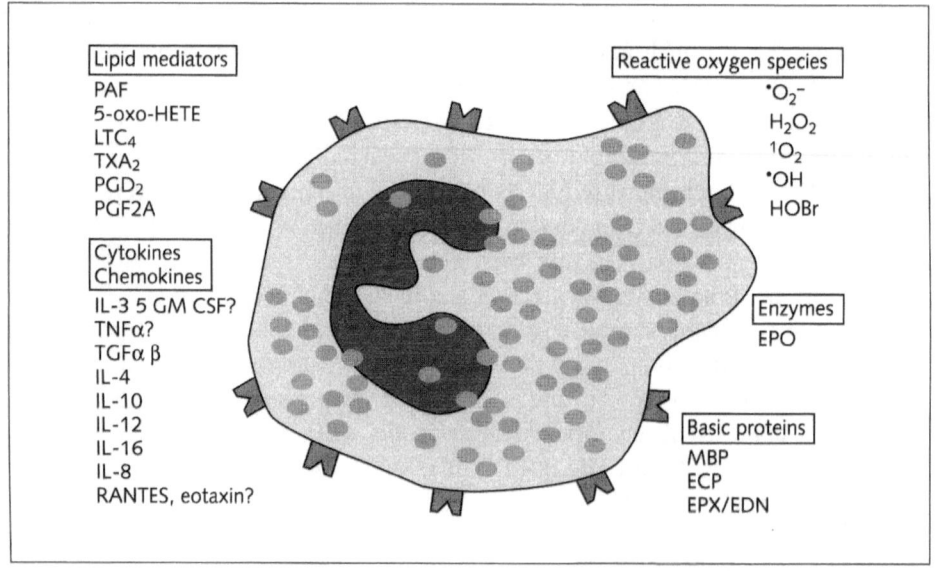

Figure 2

Synthesis and release of mediators and toxic agents.

The tissue toxic role of eosinophils is mediated by the release of reactive oxygen species, enzymes and basic proteins (right side). Eosinophils are also able to produce and release different mediators, cytokines and chemokines (left side). ? = controversial results (expressed on mRNA).

burst following stimulation with chemotaxins [54]. Prevention of $[Ca^{2+}]_i$ transients by intracellular calcium chelators resulted in a total inhibition of the release of reactive oxygen species [54]. Furthermore, this study clearly point to a synergistic effect of intracellular Ca^{2+} and protein kinase C in the activation of the respiratory burst (Fig. 3).

In another study, we could demonstrate that intracellular Ca^{2+} fluxes in eosinophils represent an important step in the activation leading to direct migratory response and actin polymerization [126]. These findings in eosinophils are in direct contrast to actin polymerization in neutrophils, in which intracellular calcium mobilization is not sufficient [127, 128]. Therefore, this study clearly shows a difference in the regulation of actin polymerization in eosinophils compared to that of neutrophils [126]. Generally, intracellular Ca^{2+} seems to play a central role in the modulation of the respiratory burst, chemotaxis and actin polymerization in eosinophils and might therefore be an interesting target for drugs which interact in the calcium homeostasis thus preventing the migration process and destructive power of eosinophils in the inflammatory tissue (Fig. 3).

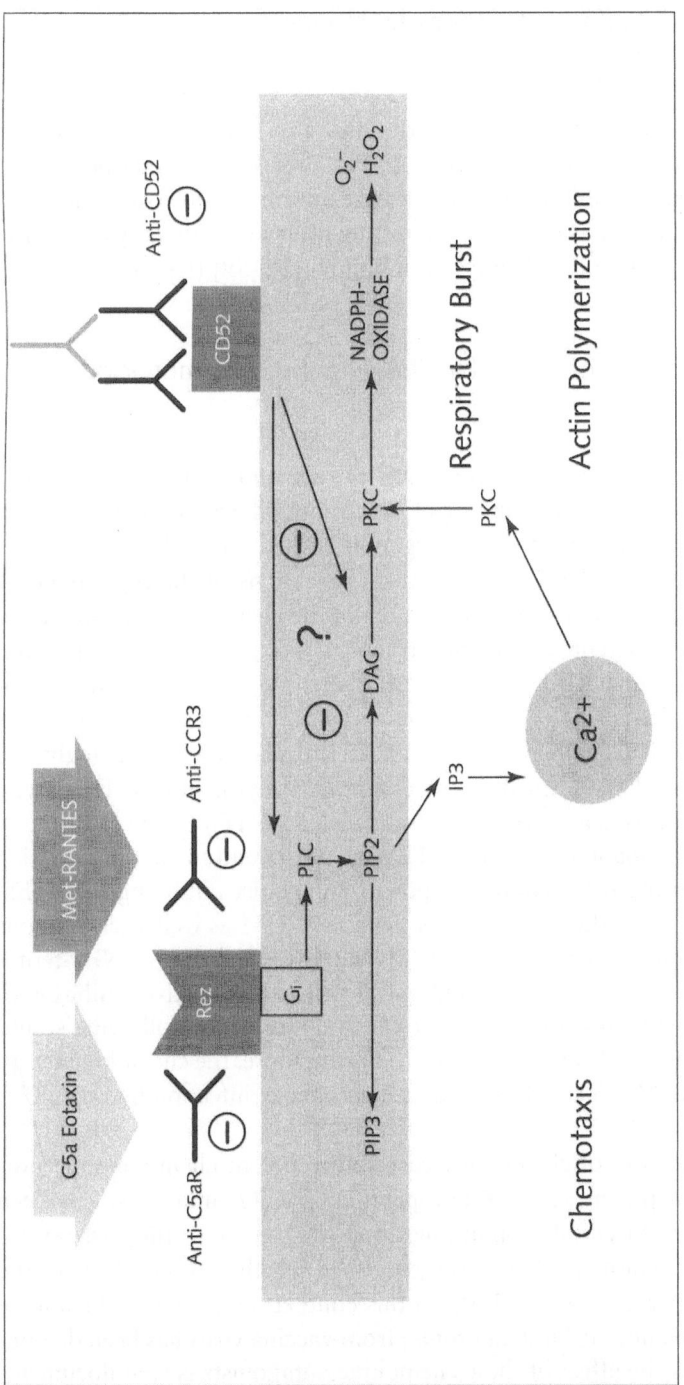

Figure 3

Modulation of eosinophil effector functions

Prevention of intracellular calcium fluxes in eosinophils results in an inhibition of chemotaxis, actin polymerization and respiratory burst. Moreover, effector mechanisms of human eosinophils are modulated by different monoclonal antibodies (against C5aR, CCR3 and CD52) or chemokine receptor antagonist (Met-RANTES). Therefore, these tools point to a potential new therapeutical approach to prevent the invasion and destructive power of eosinophils in diseases that are accompanied by eosinophil infiltration such as allergic asthma, allergic rhinitis and connective tissue diseases.

Inhibition of eosinophil effector functions by monoclonal antibodies against CD52 and CD88

The inhibitory effect of the monoclonal antibodies anti-CD88 (against the C5a receptor) and anti-CD52 (against CAMPATH-1) has already been described above. Nevertheless, based on the inhibitory effect of these antibodies on eosinophil effector functions, these antibodies seem to be interesting tools in the therapy of diseases that are accompanied by blood and tissue eosinophilia [26, 59] (Fig. 3).

Chemokine receptor antagonists and inhibitory anti-CCR3 monoclonal antibodies

Since chemokines are of major importance to attract and activate eosinophils, agents that would be able to antagonize or block eosinophil activation are of interest. It is well known that the NH_2-terminal region is critical for the biological activity of chemokines and it has been shown that modifications of the corresponding region of chemokines profoundly alter the activity of the chemokines on leukocytes [129–132]. Previously, potent chemokine receptor antagonists, such as Met-RANTES, AOP-RANTES, RANTES (*3-68*), MCP-2 (*6-76*) and others were constructed by deletion or extension of amino acids or by chemical modification on the NH_2-terminal residue of CC chemokines [130, 131, 133–138]. Interestingly, the leukocyte activation marker CD26 that possesses dipeptidyl peptidase IV activity has now been shown to truncate RANTES (*1-68*) into RANTES (*3-68*) [139].

Extension of recombinant human RANTES by the retention of the initiating methionine resulted in the production of a potent antagonist inhibiting RANTES-induced $[Ca^{2+}]_i$ transients in the promonocytic cell line THP-1 as well as chemotaxis of THP-1 cells and human T lymphocytes [133]. Moreover, deletion of NH_2-terminal residue of MCP-1, MCP-3 and RANTES resulted in a competitive inhibition of chemotaxis and enzyme release in THP-1 cell line and monocytes following stimulation with the native form of chemokines [130]. Furthermore, the chemokine receptor antagonist AOP-RANTES is able to inhibit macrophage infection by HIV [135, 136].

Besides the modification of chemokines to produce potent chemokine receptor antagonists, recently a potent selective non-peptide CXCR2 antagonist has been found to inhibit IL-8-induced neutrophil migration [80]. Also, in the past several new, virally encoded chemokines have been described which can modify both the host immune and antiviral response [140]. In this context, blockade of chemokine activity by a soluble chemokine binding protein from vaccina virus has been demonstrated [141]. Whereas the effect of these chemokine antagonists is well documented on THP-1 cell line, monocytes and lymphocytes, no data about these compounds on eosinophil effector functions exist.

We have demonstrated that Met-RANTES was able to specifically inhibit eosinophil effector functions including chemotaxis, transient $[Ca^{2+}]_i$ release, actin polymerization and release of reactive oxygen species following stimulation with RANTES, MCP-3 and eotaxin [102]. Met-RANTES seems to be able to antagonize the response of eosinophils through the CCR1 at lower concentrations and CCR3 at higher concentrations [102].

Furthermore, it has been demonstrated that the mouse anti-CCR3 mAb, 7B11, was selective for human CCR3 and able to block chemotaxis and calcium flux in human eosinophils induced by RANTES, MCP-3 and eotaxin [142]. Chemotaxis and stimulation of adhesion were abrogated completely by pretreatment of eosinophils with anti-CCR3 mAb [143]. In addition, we could demonstrate that this anti-CCR3 mAb was able to inhibit the release of reactive oxygen species following stimulation with eotaxin and eotaxin-2 [95]. Taken together, the chemokine receptor antagonist Met-RANTES and the anti-CCR3 mAb, 7B11, inhibit eosinophil but not neutrophil effector functions, indicating a more target-specific way of inhibition than other drugs such as steroid and antihistamines. Therefore, these tools point to a potential new therapeutical approach to prevent the invasion and destructive power of eosinophils in diseases that are accompanied by eosinophil infiltration such as allergic asthma, allergic rhinitis and connective tissue diseases (Fig. 3).

Acknowledgements

This work was supported by a grant from the Deutsche Forschungsgemeinschaft (EL 160/6-1).

References

1   Druilhe A, Cai Z, Haile S, Chouaib S, Pretolani M (1996) Fas-Mediated apoptosis in cultured human eosinophils. *Blood* 87: 2822–2830

2   Matsumoto K, Schleimer RP, Saito H, Iikura Y, Bochner BS (1995) Induction of apoptosis in human eosinophils by anti-fas antibody treatment *in vitro*. *Blood* 86: 1437-1443

3   Simon HU (1997) Molecular mechanisms of defective eosinophil apoptosis in diseases associated with eosinophilia. *Int Arch Allergy Immunol* 113: 206–208

4   Wedi B, Raap U, Lewrick H, Kapp A (1997) Delayed eosinophil programmed cell death *in vitro*: a common feature of inhalant allergy and extrinsic and intrinsic atopic dermatitis. *J Allergy Clin Immunol* 100: 536–543

5   Yousefi S, Hoessli DC, Blaser K, Mills GB, Simon HU (1996) Requirement of lyn and syk tyrosine kinases for the prevention of apotosis by cytokines in human eosinophils. *J Exp Med* 183: 1407–1414

6   Walsh GM, Williamson ML, Symon FA, Willars GB, Wardlaw AJ (1996) Ligation of

CD69 induces apoptosis and cell death in human eosinophils cultured with granulocyte-macrophage colony-stimulating factor. *Blood* 87: 2815–2821

7   Luttmann W, Knoechel B, Foerster M, Matthys H, Virchow Jr. JC, Kroegel C (1996) Activation of human eosinophils by IL-13. Induction of CD69 surface antigen, its relationship to messenger RNA expression, and promotion of cellular viability. *J Immunol* 157: 1678–1683

8   Simon HU, Yousefi S, Dibbert B, Levi-Schaffer F, Blaser K (1997) Anti-apoptotic signals of granulocyte-macrophage colony-stimulating factor are transduced via Jak2 tyrosine kinase in eosinophils. *Eur J Immunol* 27: 3536–3539

9   Hebestreit H, Dibbert B, Balatti I, Braun D, Schapowal A, Blaser K, Simon HU (1998) Disruption of fas receptor signaling by nitric oxide in eosinophils. *J Exp Med* 187: 415–425

10 Meagher LC, Cousin JM, Seckl JR, Haslett C (1996) Opposing effects of glucocorticoids on the rate of apoptosis in neutrophilic and eosinophilic granulocytes. *J Immunol* 156: 4422–4428

11 Czech W, Krutmann J, Budnik A, Schöpf E, Kapp A (1993) Induction of intercellular adhesion molecule 1 (ICAM-1) expression in normal human eosinophils by inflammatory cytokine. *J Invest Dermatol* 100: 417–423

12 Nagata M, Sedgwick JB, Bates ME, Kita H, Busse WW (1995) Eosinophil adhesion to vascular cell adhesion molecule-1 activates superoxide anion generation. *J Immunol* 155: 2194–2202

13 Nakayama H, Sano H, Nishimura T, Yoshidi S, Iwamoto I (1994) Role of vascular cell adhesion molecule 1/very late activation antigen 4 and intercellular adhesion molecule 1/lymphocyte function-associated antigen 1 interactions in antigen-induced eosinophil and T cell recruitment into the tissue. *J Exp Med* 179: 1145–1154

14 Schleimer RP, Sterbinsky SA, Kaiser J, Bickel CA, Klunk DA, Tomioka K, Newman W, Luscinskas FW, Gimbrone MA, Jr., McIntyre BW et al (1992) IL-4 induces adherence of human eosinophils and basophils but not neutrophils to endothelium. Association with expression of VCAM-1. *J Immunol* 148: 1086–1092

15 Springer TA (1994) Traffic Signals for lymphocyte recirculation and leukocyte emigration: The multistep paradigm. *Cell* 76: 301–314

16 Symon FA, Walsh GM, Watson SR, Wardlaw AJ (1994) Eosinophil adhesion to nasal polyp endothelium is P-selectin-dependent. *J Exp Med* 180: 371–376

17 Walsh GM, Mermod JJ, Hartnell A, Kay AB, Wardlaw AJ (1991) Human eosinophil, but not neutrophil, adherence to IL-1-stimulated human umbilical vascular endothelial cells is alpha 4 beta 1 (very late antigen-4) dependent. *J Immunol* 146: 3419–3423

18 Grangette C, Gruart V, Ouaissi MA, Rizvi F, Delespesse G, Capron A, Capron M (1989) IgE receptor on human eosinophils (FcεRII) – comparison with B Cell CD23 and association with an adhesion molecule. *J Immunol* 143: 3580–3588

19 Monteiro RC, Hostoffer RW, Cooper MD, Bonner JR, Gartland GL, Kubagawa H (1993) Definition of immunoglobulin A receptors on eosinophils and their enhanced expression in allergic individuals. *J Clin Invest* 92: 1681–1685

20 Koenderman L, Hermans SW, Capel PJ, van de Winkel JG (1993) Granulocyte-macrophage colony-stimulating factor induces sequential activation and deactivation of binding via a low-affinity IgG Fc receptor, hFc gamma RII, on human eosinophils. *Blood* 81: 2413–2419

21 Clark-Lewis I, Kim KS, Rajarathnam K, Gong JH, Dewald B, Moser B, Baggiolini M, Sykes BD (1995) Structure-activity relationships of chemokines. *J Leuk Biol* 57: 703–711

22 Combadiere C, Ahuja SK, Murphy PM (1995) Cloning and functional expression of a human eosinophil CC chemokine receptor. *J Biol Chem* 270: 16491–16494

23 Gao JL, Kuhns DB, Tiffany HL, McDermott, Li X, Francke U, Murphy PH (1993) Structure and functional expression of the human macrophage inflammatory 1a/RANTES receptor. *J Exp Med* 177: 1421–1427

24 Kitaura M, Nakajima T, Imai T, Harada S, Combadiere C, Tiffany HL, Murphy PM, Yoshie O (1996) Molecular cloning of human eotaxin, an eosinophil-selective CC chemokine, and identification of a specific eosinophil eotaxin receptor, CC chemokine receptor 3. *J Biol Chem* 271: 7725–7730

25 Neote K, DiGregorio D, Mak JY, Horuk R, Schall TJ (1993) Molecular cloning, functional expression, and signaling characteristics of a C-C chemokine receptor. *Cell* 72: 415–425

26 Elsner J, Hochstetter R, Spiekermann K, Kapp A (1996) Surface and mRNA expression of the CD52 antigen by human eosinophils but not by neutrophils. *Blood* 88: 4684–4693

27 Hartnell A, Robinson DS, Kay AB, Wardlaw AJ (1993) CD69 is expressed by human eosinophils activated *in vivo* in asthma and *in vitro* by cytokines. *Immunology* 80: 281–286

28 Nishikawa K, Morii T, Ako H, Hamada K, Saito S, Narita N (1992) *In vivo* expression of CD69 on lung eosinophils in eosinophilic pneumonia: CD69 as a possible activation marker for eosinophils. *J Allergy Clin Immunol* 90: 169–174

29 Capron A, Capron M, Grangette C, Dessaint JP (1989) IgE and inflammatory cells. *Ciba Found Symp* 147: 153–170

30 Capron M, Truong MJ, Aldebert D, Gruart V, Suemura M, Delespesse G, Tourvieille B, Capron A (1991) Heterogeneous expression of CD23 epitopes by eosinophils from patients. Relationships with IgE-mediated functions. *Eur J Immunol* 21: 2423–2429

31 Gounni AS, Lamkhioued B, Delaporte E, Dubost A, Kinet JP, Capron A, Capron M (1994) The high-affinity IgE receptor on eosinophils: From allergy to parasites or from parasites to allergy? *J Allergy Clin Immunol* 94: 1214–1216

32 Gounni AS, Lamkhioued B, Ochiai K, Tanaka Y, Delaporte E, Capron A, Kinet JP, Capron M (1994) High-affinity IgE receptor on eosinophils is involved in defence against parasites. *Nature* 367: 183–186

33 Metzger H (1992) The receptor with high affinity for IgE. *Immunol Rev* 125: 37–48

34 Bieber T, de la Salle H, Wollenberg A, Hakimi J, Chizzonite R, Ring J, Hanau D, de la

Salle C (1993) Human epidermal Langerhans cells express the high affinity receptor for immunoglobulin E (Fc epsilon RI). *J Exp Med* 175: 1285–1290

35   Terada N, Konno A, Terada Y, Fukuda S, Yamashita T, Abe T, Shimada H, Ishida K, Yoshimura K, Tanaka Y, Ra C, Ishikawa K, Togawa K (1995) IL-4 upregulates FcεRI α-chain messenger RNA in eosinophils. *J Allergy Clin Immunol* 96: 1161–1169

36   Gauchat JF, Henchoz S, Fattah D, Mazzei G, Aubry JP, Jomotte T, Dash L, Page K, Solari R, Aldebert D, Capron M, Dahinden C, Bonnefoy JY (1995) CD40 ligand is functionally expressed on human eosinophils. *Eur J Immunol* 25: 863–865

37   Ohkawara Y, Lim KG, Xing Z, Gilbetic M, Nakano K, Dolovich J, Croitoru K, Weller PF, Jordana M (1996) CD40 expression by human peripheral blood eosinophils. *J Clin Invest* 97: 1761–1766

38   Fernvik E, Hallden G, Hed J, Lundahl J (1995) Intracellular and surface distribution of CD9 in human eosinophils. *APMIS* 103: 699–706

39   Hale G, Xia MQ, Tighe HP, Dyer MJS, Waldmann H (1990) The CAMPATH-1 antigen (CDw52). *Tissue Antigens* 35: 118–127

40   Treumann A, Lifely RM, Schneider P, Ferguson MAJ (1995) Primary structure of CD52. *J Biol Chem* 270: 6088–6099

41   Xia MQ, Tone M, Packman L, Waldmann H (1991) Characterization of the CAMPATH-1 (CDw52) antigen: Biochemical analysis and cDNA cloning reveal an unusually small peptide backbone. *Eur J Immunol* 21: 1677–1684

42   Hugli TE (1989) Structure and function of C3a anaphylatoxins. *Curr Topics Microbiol Immunol* 153: 181–208

43   Takabayashi T, Vannier E, Clark BD, Margolis NH, Dinarello CA, Burke JF, Gelfand JA (1996) A new biologic role of C3a and C3a desArg – Regulation of TNF-α and IL-1β synthesis. *J Immunol* 156: 3455–3460

44   Martin U, Bock D, Arseniev L, Tornetta MA, Ames RS, Bautsch W, Köhl J, Ganser A, Klos A (1997) The human C3a receptor is expressed on neutrophils and monocytes, but not on B or T lymphocytes. *J Exp Med* 186: 199–207

45   Zwirner J, Gotze O, Moser A, Sieber A, Begemann G, Kapp A, Elsner J, Werfel T (1997) Blood- and skin-derived monocytes/macrophages respond to C3a but not to C3a (desArg) with a transient release of calcium via a pertussis toxin-sensitive signal transduction pathway. *Eur J Immunol* 27: 2317–2322

46   Nagata S, Glovsky MM, Kunkel SL (1987) Anaphylatoxin-induced neutrophil chemotaxis and aggregation. *Int Arch Allergy Appl Immunol* 82: 4–9

47   Daffern PJ, Pfeifer PH, Ember JA, Hugli TE (1995) C3a is a chemotaxin for human eosinophils but not for neutrophils. I. C3a stimulation of neutrophils is secondary to eosinophil activation. *J Exp Med* 181: 2119–2127

48   Elsner J, Oppermann M, Czech W, Kapp A (1994) C3a activates the respiratory burst in human polymorphonuclear neutrophilic leukocytes via pertussis toxin-sensitive G-proteins. *Blood* 83: 3324–3331

49   Elsner J, Oppermann M, Czech W, Dobos G, Schöpf E, Norgauer J, Kapp A (1994) C3a

activates reactive oxygen radical species production and intracellular calcium transients in human eosinophils. *Eur J Immunol* 24: 518–522

50   Oppermann M, Gotze O (1994) Plasma clearance of the human C5a anaphylatoxin by binding to leucocyte C5a receptors. *Immunology* 82: 516–521

51   Werfel T, Oppermann M, Schulze M, Krieger G, Weber M, Götze O (1992) Binding of fluorescein-labeled anaphylatoxin C5a to human peripheral blood, spleen, and bone marrow leukocytes. *Blood* 79: 152–160

52   Elsner J, Hochstetter R, Kimmig D, Kapp A (1996) Human eotaxin represents a potent activator of the respiratory burst of human eosinophils. *Eur J Immunol* 26: 1919–1925

53   Zeck Kapp G, Kroegel C, Riede UN, Kapp A (1995) Mechanisms of human eosinophil activation by complement protein C5a and platelet-activating factor: Similar functional responses are accompanied by different morphologic alterations. *Allergy* 50: 34–47

54   Elsner J, Dichmann S, Kapp A (1995) Activation of the respiratory burst in human eosinophils by chemotaxins requires intracellular calcium fluxes. *J Invest Dermatol* 105: 231–236

55   Morgan EL, Ember JA, Sanderson SD, Scholz W, Buchner R, Ye RD, Hugli TE (1993) Anti-C5a receptor antibodies: Characterization of neutralizing antibodies specific for a peptide, C5aR-(9-29), derives from the predicted amino-terminal sequence of the human C5a receptor. *J Immunol* 151: 377–388

56   Oppermann U, Raedt U, Hebell T, Götze O (1993) Probing the human receptor for C5a anaphylatoxin with site-directed antibodies. Identification of a potential ligand binding site on the NH_2-terminal domain. *J Immunol* 151: 49–55

57   Gerard NP, Gerard C (1991) The chemotactic receptor for human C5a anaphylatoxin. *Nature* 349: 614–617

58   Gerard NP, Hodges MK, Drazen JM, Weller PF, Gerard C (1989) Characterization of a receptor for C5a anaphylatoxin on human eosinophils. *J Biol Chem* 264: 1760–1766

59   Elsner J, Oppermann M, Kapp A (1996) Detection of C5a receptors on human eosinophils and inhibition of eosinophil effector functions by anti-C5a receptor (CD88) antibodies. *Eur J Immunol* 26: 1560–1564

60   Weller PF (1991) The immunobiology of eosinophils. *N Engl J Med* 324: 1110–1118

61   Zeck Kapp G, Czech W, Kapp A (1994) TNF alpha-induced activation of eosinophil oxidative metabolism and morphology – Comparison with IL-5. *Exp Dermatol* 3: 176–188

62   Kay AB, Barata L, Meng Q, Durham SR, Ying S (1997) Eosinophils and eosinophil-associated cytokines in allergic inflammation. *Int Arch Allergy Immunol* 113: 196–199

63   Karlen S, De Boer ML, Lipscombe RJ, Lutz W, Mordvinov VA, Sanderson CJ (1998) Biological and molecular characteristics of interleukin-5 and its receptor. *Int Rev Immunol* 16: 227–247

64   Kapp A, Zeck Kapp G, Czech W, Schöpf E (1994) The chemokine RANTES is more than a chemoattractant: Characterization of its effect on human eosinophil oxidative metabolism and morphology in comparison with IL-5 and GM-CSF. *J Invest Dermatol* 102: 906–914

65 Schwenk U, Schröder JM (1995) 5-oxo-eicosanoids are potent eosinophil chemotactic factors: Functional characterization and structural requirements. *J Biol Chem* 270: 15029–15036

66 Schwenk U, Morita E, Engel R, Schröder JM (1992) Identification of 5-oxo-15-hydroxy-6,8,11,13-eicosatetraenoic acid as a novel and potent human eosinophil chemotactic eicosanoid. *J Biol Chem* 267: 12482–12488

67 Baggiolini M, Dewald B, Moser B (1997) Human chemokines: an update. *Annu Rev Immunol* 15: 675–705

68 Yoshie O, Imai T, Nomiyama H (1997) Novel lymphocyte-specific CC chemokines and their receptors. *J Leukoc Biol* 62: 634–644

69 Bazan JF, Bacon KB, Hardiman G, Wang W, Soo K, Rossi D, Greaves DR, Zlotnik A, Schall TJ (1997) A new class of membrane-bound chemokine with a CX3C motif. *Nature* 385: 640–644

70 Luster AD, Rothenberg ME (1997) Role of the monocyte chemoattractant protein and eotaxin subfamily of chemokines in allergic inflammation. *J Leukoc Biol* 62: 620–633

71 Baggiolini M (1998) Chemokines and leukocyte traffic. *Nature* 392: 565–568

72 Moser B, Loetscher M, Piali L, Loetscher P (1998) Lymphocyte responses to chemokines. *Int Rev Immunol* 16: 323–344

73 Baggiolini M, Walz A, Kunkel SL (1989) Neutrophil activating peptide-1/interleukin-8, a novel cytokine that activates neutrophils. *J Clin Invest* 84: 1045–1049

74 Koch AE, Kunkel SL, Harlow LA, Mazarakis DD, Haines GK, Burdick MD, Pope RM, Walz A, Strieter RM (1994) Epithelial neutrophil activating peptide-78: a novel chemotactic cytokine for neutrophils in arthritis. *J Clin Invest* 94: 1012–1018

75 Moser B, Clark-Lewis I, Zwahlen R, Baggiolini M (1990) Neutrophil-activating properties of the melanoma growth-stimulatory activity. *J Exp Med* 171: 1792–1802

76 Schröder JM, Persoon NL, Christophers E (1990) Lipopolysaccharide stimulated human monocytes secrete apart from NAP-1/IL-8 a second neutrophil activating protein: NH_2-terminal amino acid sequence identity with melanoma growth stimulatory activity (MGSA/gro). *J Exp Med* 171: 1091

77 Schröder JM, Mrowietz U, Morita E, Christophers E (1987) Purification and partial biochemical characterization of a human monocyte-derived, neutrophil-activating peptide that lacks interleukin 1 activity. *J Immunol* 139: 3474

78 Walz A, Strieter RM, Schnyder S (1993) Neutrophil-activating peptide ENA-78. *Chemokines* 129–137

79 Ludwig A, Petersen F, Zahn S, Gotze O, Schroder JM, Flad HD, Brandt E (1997) The CXC-chemokine neutrophil-activating peptide-2 induces two distinct optima of neutrophil chemotaxis by differential interaction with interleukin-8 receptors CXCR-1 and CXCR-2. *Blood* 90: 4588–4597

80 White JR, Lee JM, Young PR, Hertzberg RP, Jurewicz AJ, Chaikin MA, Widdowson K, Foley JJ, Martin LD, Griswold DE, Sarau HM (1998) Identification of a potent, selective non-peptide CXCR2 antagonist that inhibits interleukin-8-induced neutrophil migration. *J Biol Chem* 273: 10095–10098

81   Wolf M, Delgado MB, Jones SA, Dewald B, Clark-Lewis I, Baggiolini M (1998) Granulocyte chemotactic protein 2 acts via both IL-8 receptors, CXCR1 and CXCR2. *Eur J Immunol* 28: 164–170

82   Alam R, Stafford S, Forythe P, Harrison R, Faubion D, Lett-Brown MA, Grant JA (1993) RANTES is a chemotactic and activating factor for human eosinophils. *J Immunol* 150: 3442–3447

83   Dahinden CA, Geiser T, Brunner T, von Tscharner V, Caput D, Ferrara P, Minty A, Baggiolini M (1994) Monocyte chemotactic protein 3 is a most effective basophil- and eosinophil-activating chemokine. *J Exp Med* 179: 751–756

84   Kameyoshi Y, Dörschner A, Mallet AI, Christophers E, Schröder JM (1992) Cytokine RANTES released by thrombin-stimulated platelets is a potent attractant for human eosinophils. *J Exp Med* 176: 587–592

85   Rot A, Krieger M, Brunner T, Bischoff SC, Schall TJ, Dahinden CA (1992) RANTES and macrophage inflammatory protein 1 alpha induce the migration and activation of normal human eosinophil granulocytes. *J Exp Med* 176: 1489–1495

86   Weber M, Uguccioni M, Ochsenberger B, Baggiolini M, Clark-Lewis I, Dahinden CA (1995) Monocyte chemotactic protein MCP-2 activates human basophil and eosinophil leukocytes similar to MCP-3. *J Immunol* 154: 4166–4172

87   Kodelja V, Muller C, Politz O, Hakij N, Orfanos CE, Goerdt S (1998) Alternative macrophage activation-associated CC-chemokine-1, a novel structural homologue of macrophage inflammatory protein-1 alpha with a Th2-associated expression pattern. *J Immunol* 160: 1411–1418

88   Youn BS, Zhang SM, Lee EK, Park DH, Broxmeyer HE, Murphy PM, Locati M, Pease JE, Kim KK, Antol K, Kwon BS (1997) Molecular cloning of leukotactin-1: a novel human beta-chemokine, a chemoattractant for neutrophils, monocytes, and lymphocytes, and a potent agonist at CC chemokine receptors 1 and 3. *J Immunol* 159: 5201–5205

89   Youn BS, Zhang SM, Broxmeyer HE, Cooper S, Antol K, Fraser M, Kwon BS (1998) Characterization of CKbeta8 and CKbeta8-1: two alternatively spliced forms of human beta-chemokine, chemoattractants for neutrophils, monocytes, and lymphocytes, and potent agonists at CC chemokine receptor 1. *Blood* 91: 3118–3126

90   Campbell JJ, Bowman EP, Murphy K, Youngman KR, Siani MA, Thompson DA, Wu L, Zlotnik A, Butcher EC (1998) 6-C-kine (SLC), a lymphocyte adhesion-triggering chemokine expressed by high endothelium, is an agonist for the MIP-3beta receptor CCR7. *J Cell Biol* 141: 1053–1059

91   Kim CH, Pelus LM, White JR, Applebaum E, Johanson K, Broxmeyer HE (1998) CK beta-11/macrophage inflammatory protein-3 beta/EBI1-ligand chemokine is an efficacious chemoattractant for T and B cells. *J Immunol* 160: 2418–2424

92   Tenscher K, Metzner B, Schopf E, Norgauer J, Czech W (1996) Recombinant human eotaxin induces oxygen radical production, Ca(2+)- mobilization, actin reorganization, and CD11b upregulation in human eosinophils via a pertussis toxin-sensitive heterotrimeric guanine nucleotide-binding protein. *Blood* 88: 3195–3199

93   Tenscher K, Metzner B, Hofmann C, Schopf E, Norgauer J (1997) The monocyte chemotactic protein-4 induces oxygen radical production, actin reorganization, and CD11b up-regulation via a pertussis toxin- sensitive G-protein in human eosinophils. *Biochem Biophys Res Commun* 240: 32–35

94   Petering H, Hochstetter R, Kimmig D, Smolarski R, Kapp A, Elsner J (1998) Detection of MCP-4 in dermal fibroblasts and its activation of the respiratory burst in human eosinophils. *J Immunol* 160: 555–558

95   Elsner J, Petering H, Kluthe C, Kimmig D, Smolarski R, Ponath P, Kapp A (1998) Eotaxin-2 activates chemotaxis-related events and release of reactive oxygen species via pertussis toxin-sensitive G proteins in human eosinophils. *Eur J Immunol* 28: 2152–2158

96   Ying S, Taborda Barata L, Meng Q, Humbert M, Kay AB (1995) The kinetics of allergen-induced transcription of messenger RNA for monocyte chemotactic protein-3 and RANTES in the skin of human atopic subjects: Relationship to eosinophil, T cell, and macrophage recruitment. *J Exp Med* 181: 2153–2159

97   Beck LA, Dalke S, Leiferman KM, Bickel CA, Hamilton R, Rosen H, Bochner BS, Schleimer RP (1997) Cutaneous injection of RANTES causes eosinophil recruitment: comparison of nonallergic and allergic human subjects. *J Immunol* 159: 2962–2972

98   Ben-Baruch A, Xu L, Young PR, Bengali K, Oppenheim JJ, Wang JM (1995) Monocyte chemotactic protein-3 (MCP-3) interacts with multiple leukocyte receptors – C-C CKR1, a receptor for macrophage inflammatory protei-1a /RANTES, is also a functional receptor for MCP-3. *J Biol Chem* 270: 22123–22128

99   Mackay CR (1996) Chemokine receptors and T cell chemotaxis. *J Exp Med* 184: 799–802

100 Ponath PD, Qin S, Post TW, Wang J, Wu L, Gerard NP, Newman W, Gerard C, Mackay CR (1996) Molecular cloning and characterization of a human eotaxin receptor expressed selectively on eosinophils. *J Exp Med* 183: 2437–2448

101 Daugherty BL, Siciliano SJ, DeMartino JA, Malkowitz L, Sirotina A, Springer MS (1996) Cloning, expression, and characterization of the human eosinophil eotaxin receptor. *J Exp Med* 183: 2349–2354

102 Elsner J, Petering H, Hochstetter R, Kimmig D, Wells TN, Kapp A, Proudfoot AE (1997) The CC chemokine antagonist Met-RANTES inhibits eosinophil effector functions through the chemokine receptors CCR1 and CCR3. *Eur J Immunol* 27: 2892–2898

103 Tiffany HL, Alkhatib G, Combadiere C, Berger EA, Murphy PM (1998) CC chemokine receptors 1 and 3 are differentially regulated by IL-5 during maturation of eosinophilic HL-60 cells. *J Immunol* 160: 1385–1392

104 Forssmann U, Uguccioni M, Loetscher P, Dahinden CA, Langen H, Thelen M, Baggiolini M (1997) Eotaxin-2, a novel CC chemokine that is selective for the chemokine receptor CCR3, and acts like eotaxin on human eosinophil and basophil leukocytes. *J Exp Med* 185: 2171–2176

105 White JR, Imburgia C, Dul E, Appelbaum E, O'Donnell K, O'Shannessy DJ, Brawner M, Fornwald J, Adamou J, Elshourbagy NA, Kaiser K, Foley JJ, Schmidt DB, Johanson

K, Macphee C, Moores K, McNulty D, Scott GF, Schleimer RP, Sarau HM (1997) Cloning and functional characterization of a novel human CC chemokine that binds to the CCR3 receptor and activates human eosinophils. *J Leukoc Biol* 62: 667–675

106 Garcia-Zepeda EA, Rothenberg ME, Ownbey RT, Celestin J, Leder P, Luster AD (1996) Human eotaxin is a specific chemoattractant for eosinophil cells and provides a new mechanism to explain tissue eosinophilia. *Nature Medicine* 2: 449–456

107 Ying S, Robinson DS, Meng Q, Rottman J, Kennedy R, Ringler DJ, Mackay CR, Daugherty BL, Springer MS, Durham SR, Williams TJ, Kay AB (1997) Enhanced expression of eotaxin and CCR3 mRNA and protein in atopic asthma. Association with airway hyper-responsiveness and predominant co-localization of eotaxin mRNA to bronchial epithelial and endothelial cells. *Eur J Immunol* 27: 3507–3516

108 Sallusto F, Mackay CR, Lanzavecchia A (1997) Selective expression of the eotaxin receptor CCR3 by human T helper 2 cells. *Science* 277: 2005–2007

109 Bonecchi R, Bianchi G, Bordignon PP, D'Ambrosio D, Lang R, Borsatti A, Sozzani S, Allavena P, Gray PA, Mantovani A, Sinigaglia F (1998) Differential expression of chemokine receptors and chemotactic responsiveness of type 1 T helper cells (Th1s) and Th2s. *J Exp Med* 187: 129–134

110 Qin S, Rottman JB, Myers P, Kassam N, Weinblatt M, Loetscher M, Koch AE, Moser B, Mackay CR (1998) The chemokine receptors CXCR3 and CCR5 mark subsets of T cells associated with certain inflammatory reactions. *J Clin Invest* 101: 746–754

111 Sallusto F, Lenig D, Mackay CR, Lanzavecchia A (1998) Flexible programs of chemokine receptor expression on human polarized T helper 1 and 2 lymphocytes. *J Exp Med* 187: 875–883

112 Haskell MD, Moy JN, Gleich GJ, Thomas LL (1995) Analysis of signaling events associated with activation of neutrophil superoxide anion production by eosinophil granule major basic protein. *Blood* 86: 4627–4637

113 Kita H, Abu-Ghazaleh RI, Sur S, Gleich GJ (1995) Eosinophil major basic protein induces degranulation and IL-8 production by human eosinophils. *J Immunol* 154: 4749–4758

114 Lim KG, Wan HC, Bozza PT, Resnick MB, Wong DTW, Cruikshank WW, Kornfeld H, Center DM, Weller PF (1996) Human eosinophils elaborate the lymphocyte chemoattractants – IL-16 (lymphocyte chemoattractant factor) and RANTES. *J Immunol* 156: 2566–2570

115 Yousefi S, Hemmann S, Weber M, Hölzer C, Hartung K, Blaser K, Simon HU (1995) IL-8 is expressed by human peripheral blood eosinophils – Evidence for increased secretion in asthma. *J Immunol* 154: 5481–5490

116 Ying S, Meng Q, Taborda Barata L, Corrigan CJ, Barkans J, Assoufi B, Moqbel R, Durham SR, Kay AB (1996) Human eosinophils express messenger RNA encoding RANTES and store and release biologically active RANTES protein. *Eur J Immunol* 26: 70–76

117 Lacy P, Levi-Schaffer F, Mahmudi-Azer S, Bablitz B, Hagen SC, Velazquez J, Kay AB,

Moqbel R (1998) Intracellular localization of interleukin-6 in eosinophils from atopic asthmatics and effects of interferon gamma. *Blood* 91: 2508–2516

118 Levi-Schaffer F, Weg VB (1997) Mast cells, eosinophils and fibrosis. *Clin Exp Allergy* 27 (1): 64–70

119 Varga J, Kahari VM (1997) Eosinophilia-myalgia syndrome, eosinophilic fasciitis, and related fibrosing disorders. *Curr Opin Rheumatol* 9: 562–570

120 Grewe M, Czech W, Morita A, Werfel T, Klammer M, Kapp A, Ruzicka T, Schopf E, Krutmann J (1998) Human eosinophils produce biologically active IL-12: implications for control of T cell responses. *J Immunol* 161: 415–420

121 Maggi E (1998) The TH1/TH2 paradigm in allergy. *Immunotechnology* 3: 233–244

122 Berridge MJ (1993) Inositol triphosphate and calcium signalling. *Nature* 361: 315–325

123 Tsien RW, Tsien RY (1990) Calcium channels, stores, and oscillations. *Ann Rev Cell Biol* 6: 715–760

124 Hendey B, Lawson M, Marcantonio EE, Maxfield FR (1996) Intracellular calcium and calcineurin regulate neutrophil motility on vitronectin through a receptor identified by antibodies to integrins av and b3. *Blood* 87: 2038–2048

125 Kernen P, Wymann MP, von Tscharner V, Deranleau DA, Tai PC, Spry CJ, Dahinden CA, Baggiolini M (1991) Shape changes, exocytosis, and cytosolic free calcium changes in stimulated eosinophils. *J Clin Invest* 87: 2012–2017

126 Elsner J, Dichmann S, Dobos GJ, Kapp A (1996) Actin polymerization in human eosinophils, unlike human neutrophils, depends on intracellular calcium mobilization. *J Cell Physiol* 167: 548–555

127 Downey GP, Chan CK, Trudel S, Grinstein S (1990) Actin assembly in electropermeabilized neutrophils: Role of intracellular calcium. *J Cell Biol* 110: 1975–1982

128 Salmon JE, Brogle NL, Edberg JC, Kimberly RP (1991) Fc receptor III induces actin polymerization in human neutrophils and primes phagocytosis mediated by Fcg receptor II. *J Immunol* 146: 997–1004

129 Gong JH, Clark-Lewis I (1995) Antagonists of monocyte chemoattractant protein 1 identified by modification of functionallly critical NH$_2$-terminall residues. *J Exp Med* 181: 631–640

130 Weber M, Uguccioni M, Baggiolini M, Clark Lewis I, Dahinden CA (1996) Deletion of the NH$_2$-terminal residue converts monocyte chemotactic protein 1 from an activator of basophil mediator release to an eosinophil chemoattractant. *J Exp Med* 183: 681–685

131 Wells TNC, Power CA, Lusti-Narasimhan M, Hoogewerf AJ, Cooke RM, Chung C, Peitsch MC, Proudfoot AEI (1996) Selectivity and antagonism of chemokine receptors. *J Leukoc Biol* 59: 53–60

132 Baggiolini M, Moser B (1997) Blocking chemokine receptors. *J Exp Med* 186: 1189–1191

133 Proudfoot AE, Power CA, Hoogewerf AJ, Montjovent MO, Borlat F, Offord RE, Wells TN (1996) Extension of recombinant human RANTES by the retention of the initiating methionine produces a potent antagonist. *J Biol Chem* 271: 2599–2603

134 Teixeira MM, Wells TN, Lukacs NW, Proudfoot AE, Kunkel SL, Williams TJ, Hellewell

PG (1997) Chemokine-induced eosinophil recruitment. Evidence of a role for endogenous eotaxin in an *in vivo* allergy model in mouse skin. *J Clin Invest* 100: 1657–1666

135 Simmons G, Clapham PR, Picard L, Offord RE, Rosenkilde MM, Schwartz TW, Buser R, Wells TNC, Proudfoot AE (1997) Potent inhibition of HIV-1 infectivity in macrophages and lymphocytes by a novel CCR5 antagonist. *Science* 276: 276–279

136 Mack M, Luckow B, Nelson PJ, Cihak J, Simmons G, Clapham PR, Signoret N, Marsh M, Stangassinger M, Borlat F, Wells TNC, Schlondorff D, Proudfoot AEI (1998) Aminooxypentane-RANTES induces CCR5 internalization but inhibits recycling: A novel inhibitory mechanism of HIV infectivity. *J Exp Med* 187: 1215–1224

137 Struyf S, De Meester I, Scharpe S, Lenaerts JP, Menten P, Wang JM, Proost P, Van Damme J (1998) Natural truncation of RANTES abolishes signaling through the CC chemokine receptors CCR1 and CCR3, impairs its chemotactic potency and generates a CC chemokine inhibitor. *Eur J Immunol* 28: 1262–1271

138 Proost P, Struyf S, Couvreur M, Lenaerts JP, Conings R, Menten P, Verhaert P, Wuyts A, Van Damme J (1998) Posttranslational modifications affect the activity of the human monocyte chemotactic proteins MCP-1 and MCP-2: identification of MCP- 2(6-76) as a natural chemokine inhibitor. *J Immunol* 160: 4034–4041

139 Oravecz T, Pall M, Roderiquez G, Gorrell MD, Ditto M, Nguyen NY, Boykins R, Unsworth E, Norcross MA (1997) Regulation of the receptor specificity and function of the chemokine RANTES (regulated on activation, normal T cell expressed and secreted) by dipeptidyl peptidase IV (CD26)-mediated cleavage. *J Exp Med* 186: 1865–1872

140 Wells TN, Schwartz TW (1997) Plagiarism of the host immune system: lessons about chemokine immunology from viruses. *Curr Opin Biotechnol* 8: 741–748

141 Alcami A, Symons JA, Collins PD, Williams TJ, Smith GL (1998) Blockade of chemokine activity by a soluble chemokine binding protein from vaccinia virus. *J Immunol* 160: 624–633

142 Heath H, Qin S, Rao P, Wu L, LaRosa G, Kassam N, Ponath PD, Mackay CR (1997) Chemokine receptor usage by human eosinophils. The importance of CCR3 demonstrated using an antagonistic monoclonal antibody. *J Clin Invest* 99: 178–184

143 Kitayama J, Mackay CR, Ponath PD, Springer TA (1998) The C-C chemokine receptor CCR3 participates in stimulation of eosinophil arrest on inflammatory endothelium in shear flow. *J Clin Invest* 101: 2017–2024

Mast cells

Edward F. Knol

Department of Dermatology/Allergology, G02-124, University Medical Center Utrecht,
Heidelberglaan 100, NL-3584 CX Utrecht, The Netherlands

Introduction

During his studies in Leipzig, more than 100 years ago, Paul Ehrlich described several tissue-bound and blood leukocytes according to their staining properties with basic and azurophilic dyes [1]. One particular cell type in the tissue contained many cytoplasmic granules which stained with a high affinity for certain basic dyes. These cells were located in the connective tissue near blood vessels. The fact that these cells increased in number when the frogs that were studied were put on a highly nutritious diet, and the cells appeared "overfed", caused Ehrlich to name these "mast cells", derived from the German word "Mästung" [2].

Later studies have demonstrated that mast cells are also located in mucosal tissues. In general, it is clear that mast cells are especially located at these tissue sites where contact with the environment can occur, such as lung, skin, nose and intestines [3].

Mast cells have a cell size ranging between 8 and 20 μm, They contain numerous cytoplasmic granules and lipid bodies, and express on their cell surface the high affinity receptor for IgE (FcεRI), which is usually occupied by IgE.

Mast cell differentiation

Mast cells are derived from immature presursor cells arising in the bone marrow [3]. Although normally bone marrow is the most likely site for the origin of the mast cells progenitors, mast cells have also been cultured from progenitors obtained from fetal liver and umbilical cord blood [4, 5]. The mast cell presursor cells are thought to enter their target organ and acquire their final phenotype under the influence of local resident cells. *In vitro* maturation of mast cells from progenitor cells demonstrated the requirement of fibroblasts, secreting stem cell factor (SCF), also named C-kit-ligand [6]. Optimal maturation into mast cells from cord blood progenitors is induced by both SCF and interleukin-6 [7]. These *in vitro* differentiated mast cells

Immunology and Drug Therapy of Allergic Skin Diseases, edited by C.A.F.M. Bruijnzeel-Koomen and E.F. Knol

contain both tryptase and histamine, have FcεRI on their membrane, as well as other surface antigens typical for mast cells, and degranulate after FcεRI cross-linking. The Th2 cytokine IL-4 inhibits maturation and proliferation of immature human mast cells [8].

The hematopoietic lineage giving rise to human mast cells is at present not known. Previously, it was thought that mast cells were derived from basophils, like macrophages are derived from monocytes. Mast cells were therefore sometimes referred to as tissue basophils. With the availability of purification methods for leukocytes, specific antibodies and recombinant cytokines, it became clear that mast cells do not arise from basophils. Also, mast cells were thought to differentiate from a monocytes fraction [9]. Later studies, using more sophistigated antibody-based separation techniques, demonstrated that mast cells could only be derived from CD34+ stem cells that are present in the blood by the induction with SCF. No mast cell differentiation could be induced by stimulating highly purified lymphocytes, basophils or monocytes [10]. Extensive flow cytometric analysis, using well-defined CD antigens (CD1-130), demonstrated a poor phenotypic relationship between mast cells, basophils and monocytes [11]. This implicates that mast cells have their own hematopoietic, myeloid lineage.

Mast cell products

Upon activation, mast cells release many pro-inflammatory mediators and cytokines [12]. In general, two different release mechanisms are used. First is the relative rapid release of stored mediators *via* degranulation. This results in release of, for instance, histamine and tryptase leading to smooth muscle contraction, vasodilation and increased vasopermeability. Moreover, the mast cell contains the cytokine tumor-necrosis factor-α (TNFα) within its granules, which is released upon degranulation [13]. TNFα is an important activator of endothelium and epithelium and induces the recruitment of leukocytes [12]. Mast cells are unique cells in that they do store relatively high levels of this cytokine. The second mechanism is *via* the release of *de novo* synthesized lipid mediators and cytokines. The lipid mediators released by mast cells, such as prostaglandins, leukotrienes and platelet-activating factor are very potent at the molecular level in inducing bronchoconstriction, increased vasopermeability, leukocyte activation and attraction [12]. The newly synthesized cytokines are released at timepoints later than the preformed and *de novo* synthesized mediators and their effect has a slower onset, but is more prolonged and is considered to be especially important in the late phase of the allergic response. Cytokines released by human mast cells, such as IL-4 and IL-5 are important in activation and attraction of basophils and eosinophils, but do also promote synthesis of IgE by B cells and specific maturation of eosinophils [14]. An extensive overview of the products released by mast cells is given in Table 1.

Table 1 - Human mast cell products [3, 12, 34]

Mediators	Effect
Preformed	
histamine	vasodilatation
	increased vasopermeability
	increased fluid secretion
	smooth muscle contraction
heparin	anticoagulant
	activation of tryptase
serotonin	vasodilatation
	increased vasopermeability
chymase	increased mucus secretion
	smooth muscle contration
	cleaves neuropeptides
tryptase	activation of fibroblast
	cleavage of C3 and C3a
	degradation of neuropeptides
TNFα	recruitment of leukocytes
	activation of leukocytes
	activation of endothelium and epithelium
newly synthesized	
leukotriene C4	increased mucus secretion
	smooth muscle contraction
	bronchoconstriction
	increased vasopermeability
leukotriene B4	increased vasopermeability
	recruitment of leukocytes
	activation of leukocytes
prostaglandin D2	increased fluid secretion
	increased mucus secretion
	bronchoconstriction
platelet-activating factor	activation of leukocytes
	activation of epithelium
	platelet activation
	increased vasopermeability
	bronchoconstriction
interleukin-4	activation of endothelium
	activation of leukocytes
	promotion IgE synthesis

Table 1 (continued)

Mediators	Effect
interleukin-5	activation of basophils and eosinophils
	differentiation of eosinophils
interleukin-6	production of acute-phase proteins
	activation of leukocytes
	increased immunoglobulin synthesis
interleukin-13	activation of endothelium
	activation of leukocytes
	promotion of IgE synthesis
GM-CSF	activation of leukocytes

Mast cell activation

Although a variety of agents can initiate mast cell degranulation, the best-studied pathway of stimulation is that transduced through the high affinity IgE receptor (FcεRI) expressed on the mast cells surface [15]. This process occurs without evidence of toxicity to the cell. Moreover, mast cells recover after degranulation [16]. Because mast cells are long-lived cells, they can thus potentially participate in multiple rounds of FcεRI-dependent activation.

IgE receptor-mediated stimulation of mast cells

Cross-linking of IgE bound to the high-affinity receptor for IgE (FcεRI) (Ka = 10^{10} mol/L) gives rise to mast cell degranulation [17]. Mast cells express 10^4 to 10^6 FcεRI per cell on their surface [18]. The number of FcεRI is regulated, at least in part, by the concentration of IgE [19]. Although crosslinking of IgE by allergens is the most important mechanism for mast cells in atopic diseases, also other agents can crosslink FcεRI. A subgroup of patients with chronic urticaria have auto-antibodies to FcεRI and have stimulated skin mast cells due to the binding of these antibodies [20]. Several mitogens, such as concanavalin A, or phytohemagglution bind to the carbohydrates on the IgE molecules and can cause histamine release. Also, certain bacterial proteins, such as protein A from *Staphylococcus aureus* bind to IgE and can, in this non-specific manner, cause mast cell degranulation [15].

The FcεRI consists of four subunits, one α-chain, one β-chain and two γ-chains. The α-chain is responsible for the binding of IgE, but appears to be of no importance to the signalling [21]. Also the β-chain seems to be not essential for activation *via* the IgE receptor [22]. Associated with the β-chain is a tyrosine kinase of the src family, Lyn. Like the β-chain, the two γ-chains possess so-called ITAMs (immunoreceptor tyrosine-based activation motifs) that are indispensable for signalling *via* the IgE receptor. These ITAMs are conserved sequences that are found also in elements of the T cell-, of the B cell- and of the Fcγ-receptors antigen receptor [23]. The ITAMs on the γ-chains of the FcεRI are docking sites for the SH2 (Src homology region 2) domains of a non-transmembrane protein tyrosine-kinase, Syk.

Signal transduction starts with clustering of the IgE receptor, leading to (auto)phosphorylation of the β-chain and Lyn [24]. Subsequently, the ITAMs in the γ-chains are phosphorylated. This reveals the docking sites for Syk. Syk is transferred to the plasma membrane, where it is phosphorylated. Although this process also takes place in the absence of Lyn, phosphorylation of the γ-chains as well as of Syk is potently enhanced by Lyn, thus amplifying the reaction [25]. The Fcε receptor is also expressed in cells other than basophils or mast cells [26–29] lack this amplification because the receptor in these cells lack the β-chain. Activated Syk phosphorylates phospholipase C, which causes turnover of phosphatidylinositol-4,5-bisphosphate (PIP2) to diacylglycerol and inositol-1,4,5-trisphosphate (IP3). Subsequently, the free cytosolic Ca^{2+} concentration is increased by Ca^{2+} release from IP3-sensitive, intracellular Ca^{2+} stores. Furthermore, diacylglycerol enhances the activity of PKC. These occurrences cause reorganisation of the cytoskeleton and lead to degranulation. Activation of syk also results *via* the Ras/MAP kinase pathway and causes activation of membrane phospholipids. Phospholipase A2 then cleaves phosphocholine, resulting in formation of arachidonic acid. Moreover, the phospholipase A2 can by hydrolysis of phosphocholine result in lyso-PAF which is further acetylated into platelet-activating factor (PAF). Arachidonic acid can be converted further by the lipoxygenase pathway in leukotriene C4, whereas the cyclooxygenase pathway converts arachidonic acid into prostaglandin D2. The signal-transduction mechanisms involved in FcεRI-mediated mast cell activation are depicted in Figure 1.

Non IgE receptor-mediated stimulation of human mast cells

Besides *via* their IgE receptors, mast cells can be stimulated by a variety of both immunological and non-immunological stimuli. The complement anaphylatoxins C5a and C3a can stimulate mast cells, of which C5a is the most potent [30]. Binding of CD88 antibodies recognizing the C5a receptor has been demonstrated on skin mast cells [31]. Stem cell factor is not only involved in the development of mast cells, but also in the stimulation of mast cells leading to degranulation [32]. Moreover, SCF is important in the chemotactic movement of mast cells [33].

Non-immunological stimuli include morphine, codeine, muscle relaxants and the histamine releaser compound 48/80 [34]. Also the neuropeptides substance P, vaso-

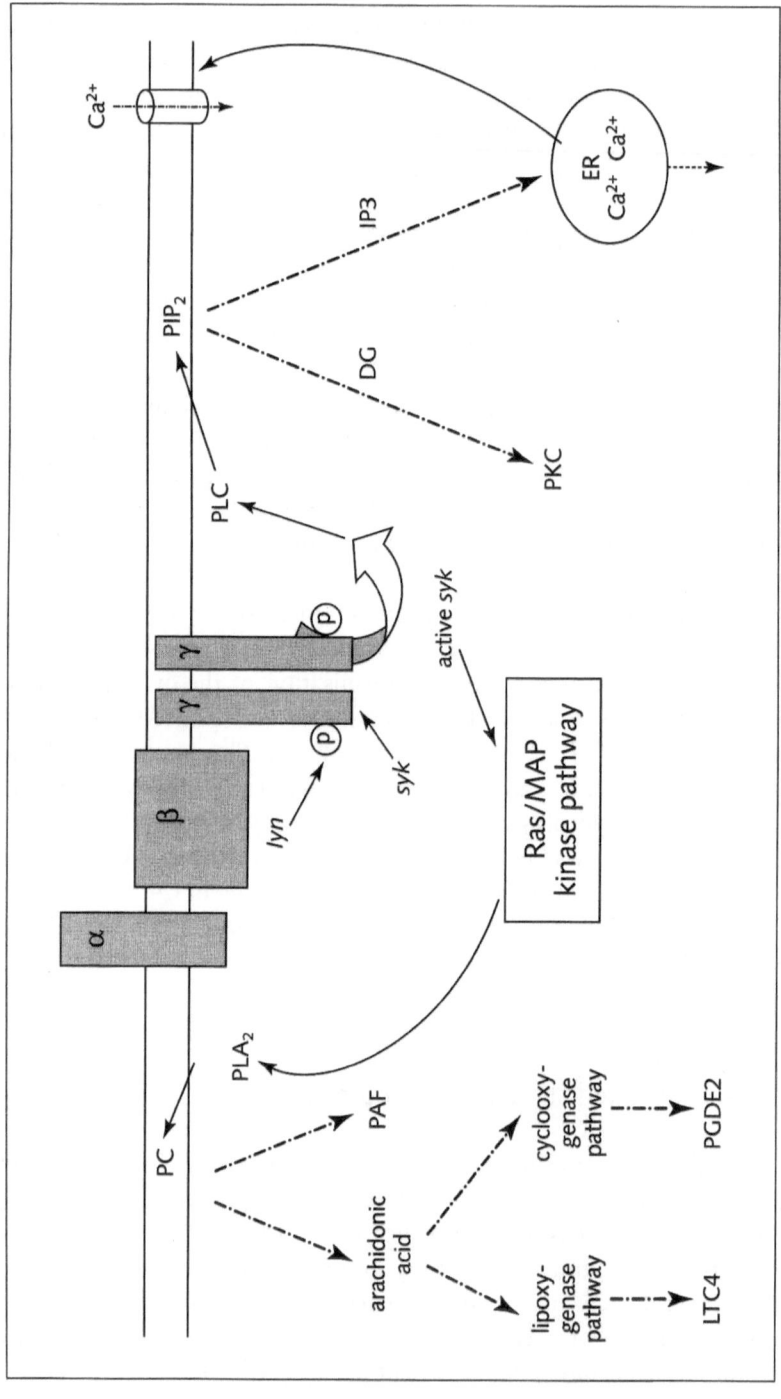

Figure 1

Signal transduction pathways of mast cells through FcεRI. For explanation see text. DG, diacylglycerol; ER, endoplasmic reticulum; IP3, inositol-1,4,5-trisphosphate; PAF, platelet-activating factor; PC, phosphocholine; PKC, protein kinase C; PLA₂, phospholipase A₂; PLC, phospholipase C.

active intestinal polypeptide and somatostatin do induce mast cell degranulation [35]. Whereas, as already described by Ehrlich, mast cells have been found in close proximity to nerve endings [2], it is suggestive that the nerve system can directly activate mast cells resulting in a local inflammatory response.

Recently, interactions have been demonstrated between activated T lymphocytes and mast cells resulting in mast cell degranulation [36–38]. This effect was found to be dependent on cell-to-cell contact and was mediated, at least in part, by β2 integrins and ICAM-1.

In murine models it has been shown that direct activation by bacterial compounds results in the release of TNFα from mast cells which is of crucial importance for the host defense against bacteria [39, 40]. This implicates that mast cells play an important role in both the natural and the acquired immunity. *Ergo*, mast cells are not only detrimental cells with regard to allergic inflammatory reaction, but also may have an important role in the natural immunity against bacteria [41].

Effect of mast cells by cell-to-cell contact

Mast cells not only activate other cells *via* the release of pro-inflammatory mediators and cytokines, but can stimulate other cells *via* direct cell-to-cell contact. Mast cells directly stimulate IgE synthesis by B cells by the expression of CD40L on the mast cells binding to CD40 on the B cells [42]. This IgE synthesis was even more pronounced when the mast cells were stimulated by allergen to release IL-4 and IL-13 [43, 44].

After stimulation with GM-CSF murine bone marrow-derived mast cells express the co-stimulatory molecules CD80 and CD86. *Via* this co-stimulatory pathway T lymphocytes are activated. This activation can be blocked by addition of mCTLA-4 [45]. Although it has been demonstrated that mast cells possess antigen-presenting properties and can transport antigen from the skin into the draining lymph nodes [46–49], it is still under debate if this is of any physiological meaning, as compared to the antigen-processing and presentation capacity of dendritic cells.

With regard to the effect of human mast cells on human T lymphocytes, it was described that activation of human mast cells resulted in modulation of CD8+ T lymphocytes function and increased IFNγ production by these T lymphocytes [50]. Which factors, soluble or membrane-bound, are involved and whether this is limited to the CD8+ subsets of T lymphocytes is not known yet.

Mast cells in skin

In atopic dermatitis only modest, if any, increase in skin mast cell number compared to controls is found [51]. Repeated allergen application to mildly abraded skin resulted in increased mast cell numbers in skin [52]. A more significant increase in mast cell numbers has been described for chronic urticaria (10 times increased compared to normal skin) [53]. However, the most dramatic increase in skin mast cell number is demonstrated in systemic mastocytosis [54].

Mast cells in skin have a special phenotype compared to other tissues. Originally, in mouse two distinct mast cell subtypes were described based on their contents. One subtype was present in mucosal tissue and contained tryptase but no chymase (MCt), whereas the other subtype contained both tryptase and chymase and was found in connective tissue (MCtc) [55]. This division is less pronounced in human tissue. Nasal mucosa, intestinal mucosa, bronchi and conjunctiva contain both types of mast cells. Only skin, synovium and the vascular wall contain mast cells restricted to the MCtc lineages, whereas the alveolar wall contains predominantly MCt mast cells [3]. Mast cells from skin are considered to be the most responsive mast cell type. In contrast to other tissue mast cells, they respond to stimulation with the neuropeptide substance P, complement fragments C3a and C5a and stem cell factor [34]. Also on the basis of pharmacological modulation, it is demonstrated that skin mast cells represent a unique population. For instance, it has been shown that the so-called mast cell "stabilizers" sodium cromoglycate and nedocromil sodium are potent inhibitors of IgE-mediated activation of human tonsil and lung mast cells and mast cells cultured from human cord blood stem cells, but are ineffective on mast cells derived from human skin [56]. Moreover, both drugs induced tachyphyaxis in lung and tonsil mast cells, but this was not found with skin mast cells [56].

The role of mast cells in atopic dermatitis

Although degranulation of mast cells induced by allergen exposure is obvious in nose and lung, this is less pronounced in skin. The most likely explanation is the location of the mast cells in skin; mast cells are located in the dermis and an allergen must first cross the epidermis before contacting mast cells, whereas in nose and lung mast cells are located within the epithelium. In human model systems of allergic inflammation it is found that intracutaneous allergen injection in skin results in mast cells degranulation and infiltration of inflammatory leukocytes [57]. However, this cellular infiltrate does not correlate with the characteristics of the cellular infiltrate that is found in lesional skin. A more related cellular infiltrate is induced by the atopy-patch where an allergen is applied on the skin, however mast cells degranulation is not clear in this model [57], suggesting that "full blown" mast cell degranulation is not involved in the pathogenesis of atopic dermatitis. In addition antihistamines are not effective, based on their blocking of local histamine receptors in skin [58], although they might exert some effect due to their sedative, or anti-inflammatory properties.

Although mast cells degranulation and release of histamine is probably not involved in the pathogenesis of AD, several studies indicate an important role of mast cells cytokines in the induction of inflammatory infiltrates [59]. Mast cells release of TNFα relates to the increased expression of the adhesion molecules ICAM-1 and E-selectin on skin endothelium [60, 61]. Moreover, mast cells have been reported to be the major source of IL-4 in skin, although this has not been reproduced by others [62, 63].

Analysis of mast cells colocalization with T lymphocytes in AD skin reveals tight interactions. Whether mast cells activate these T cells *via* their cytokines, or cell-to-cell contact is likely, but has not been proven.

References

1   Ehrlich P (1878) *Beiträge zur Theorie und Praxis der histologische Färbung.* Thesis. University of Leipzig

2   Ehrlich P (1879) Beiträge zur Kenntnis der granulierten Zellen u.s.w. *Verhandlungen der physiol Gesellschaft Berlin* 21: 53–60

3   Schwartz LB, Huff TF (1998) Biology of mast cells. In: E Middleton, EF Ellis, JW Yunginger, CE Reed, NF Adkinson, WW Busse (eds): *Allergy. Principles and practice.* Mosby Year Book, St. Louis, 261–276

4   Irani A-MA, Nilsson G, Miettinen U, Craig SS, Ashman LK, Ishizaka T, Zsebo KM, Schwartz LB (1992) Recombinant human stem cell factor stimulates differentiation of mast cells from dispersed human fetal liver cells. *Blood* 80: 3009–3021

5   Mitsui, H, Furitsu T, Dvorak AM, Irani AM, Schwartz LB, Inagaki N, Takei M, Ishizaka K, Zsebo MK, Gillis S et al (1993) Development of human mast cells from umbilical cord blood cells by recombinant human and murine c-kit ligand. *Proc Natl Acad Sci USA* 90: 735–739

6   Levi-Schaffer F, Rubinchik E (1994) Mast cell/fibroblast interactions. *Clin Exp Allergy* 24: 1016–1021

7   Saito H, Ebisawa M, Tachimoto H, Shichijo M, Fukagawa K, Matsumoto K, Iikura Y, Awaji T, Tsujimoto G, Yanagida M et al (1996) Selective growth of human mast cells induced by Steel factor, IL-6, and prostaglandin E2 from cord blood mononuclear cells. *J Immunol* 157: 343–350

8   Nilsson G, Miettinen U, Ishizaka T, Ashman LK, IraniA-M, Schwartz LB (1994) Interleukin-4 inhibits the expression of Kit and tryptase during stem cell factor-dependent development of human mast cells from fetal liver cells. *Blood* 84: 1519–1527

9   Czarnetzki BM, Figdor CG, Kolde G, Vroom T, Aalberse RC, de Vries JE (1984) Development of human connective tissue mast cells from purified blood monocytes. *Immunology* 51: 549–554

10 Agis H, Willheim M, Sperr WR, Wilfing A, Krömer E, Kabrna E, Spanblöchl E, Strobl H, Geissler K, Spittler A et al (1993) Monocytes do not make mast cells when cultured

in the presence of SCF: Characterization of the circulating mast cell progenitor as a c-kit⁺, CD34⁺, Ly⁻, CD14⁻, CD17⁻, colony-forming cell. *J Immunol* 151: 4221–4227

11   Agis H, Fureder W, Bankl HC, Kundi M, Sperr WR, Willheim M, Boltz Nitulescu G, Butterfield JH, Kishi K, Lechner K, Valent P (1996) Comparative immunophenotypic analysis of human mast cells, blood basophils and monocytes. *Immunology* 87: 535–543

12   Church MK, Holgate ST, Shute JK, Walls AF, Sampson AP (1998) Mast cell-derived mediators. In: E Middleton, EF Ellis, JW Yunginger, CE Reed, NF Adkinson, WW Busse, (eds): *Allergy. Principles and practice.* Mosby Year Book, St. Louis, 146–182

13   Gordon JR, Galli SJ (1990) Mast cells as a source of both preformed and immunologically inducible TNF-α/cachectin. *Nature* 346: 274–276

14   Bradding P (1996) Human mast cell cytokines. *Clin Exp Allergy* 26: 13–19

15   Siraganian RP (1998) Biochemical event in basophil or mast cell activation and mediator release. In: E Middleton, CE Reed, EF Ellis, NF Adkinson, JW Yunginger, WW Busse (eds): *Allergy Principles and practice.* Mosby Year Book, St. Louis, 204–227

16   Dvorak AM, Schleimer RP, Schulman ES, Lichtenstein LM (1986) Human mast cells use conservation and condensation mechanisms during recovery from degranulatio. *In vitro* studies with mast cells purified from human lungs. *Lab Invest* 54: 663–678

17   Metzger H (1991) The high affinity receptor for IgE on mast cells. *Clin Exp Allergy* 21: 269–279

18   Coleman JW, Godfrey RC (1981) The number and affinity of IgE receptors on dispersed human lung mast cells. *Immunology* 44: 859–863

19   MacGlashan DW Jr, Bochner BS, Adelman DC, Jardieu PM, Togias A, Lichtenstein LM (1997) Serum IgE level drives basophil and mast cell IgE receptor display. *Int Arch Allergy Immunol* 113: 45–47

20   Hide M, Francis DM, Grattan CEH, Hakimi J, Kochan JP, Greaves MW (1993) Autoantibodies against the high-affinity IgE receptor as a cause of histamine release in chronic urticaria. *N Engl J Med* 328: 1599–1604

21   Alber G, Miller L, Jelsema CL, Varin-Blank N, Metzger H (1991) Structure-function relationships in the mast cell high affinity receptor for IgE. Role of the cytoplasmic domains and of the beta subunit. *J Biol Chem* 266: 22613–22620

22   Rivera VM, Brugge JS (1995) Clustering of Syk is sufficient to induce tyrosine phosphorylation and release of allergic mediators from rat basophilic leukemia cells. *Mol Cell Biol* 15: 1582–1590

23   Vivier E, Daëron M (1997) Immunoreceptor tyrosine-based inhibition motifs. *Immunol Today* 18: 286–291

24   Scharenberg AM, Kinet J-P (1994) Initial events in FcεRI signal transduction. *J Allergy Clin Immunol* 94 Suppl: 1142–1146

25   Lin S, Cicala C, Scharenberg AM, Kinet J (1996) The FcεRIbeta subunit functions as an amplifier of FcεRIgamma-mediated cell activation signals. *Cell* 85: 985–995

26   Gounni AS, Lamkhioued B, Ochiai K, Tanaka Y, Delaporte E, Capron A, Kinet J, Capron M (1994) High-affinity IgE receptor on eosinophils is involved in defence

against parasites. *Nature* 367: 183–186

27   Wang B, Rieger A, Kilgus O, Ochiai K, Maurer D, Fodinger D, Kinet JP, Stingl G (1992) Epidermal Langerhans cells from normal human skin bind monomeric IgE via Fc epsilon RI. *J Exp Med* 175: 1353–1365

28   Maurer D, Fiebiger S, Ebner C, Reininger B, Fischer GF, Wichlas S, Jouvin MH, Schmitt Egenolf M, Kraft D, Kinet JP, Stingl G (1996) Peripheral blood dendritic cells express Fc epsilon RI as a complex composed of Fc epsilon RI alpha- and Fc epsilon RI gamma-chains and can use this receptor for IgE-mediated allergen presentation. *J Immunol* 157: 607–616

29   Maurer D, Fiebiger E, Reininger B, Wolff Winiski B, Jouvin MH, Kilgus O, Kinet JP, Stingl G (1994) Expression of functional high affinity immunoglobulin E receptors (Fc epsilon RI) on monocytes of atopic individuals. *J Exp Med* 179: 745–750

30   El-Lati SG, Dahinden CA, Church MK (1994) Complement peptides C3a- and C5a-induced mediator release from dissociated human skin mast cells. *J Invest Dermatol* 102: 803–806

31   Ghannadan M, Baghestanian M, Wimazal F, Eisenmenger M, Latal D, Kargul G, Walchshofer S, Sillaber C, Lechner K, Valent P (1998) Phenotypic characterization of human skin mast cells by combined staining with toluidine blue and CD antibodies. *J Invest Dermatol* 111: 689–695

32   Columbo M, Horowitz EM, Botana LM, MacGlashan DW Jr, Bochner BS, Gillis S, Zsebo KM, Galli SJ, Lichtenstein LM (1992) Effect of recombinant human c-kit receptor ligand on mediator release from human skin mast cells. *Int Arch Allergy Immunol* 99: 323–325

33   Nilsson G, Butterfield JH, Nilsson K, Siegbahn A (1994) Stem cell factor is a chemotactic factor for human mast cells. *J Immunol* 153: 3717–3723

34   Church MK, Bradding P, Walls AF, Okayama Y (1997) Human mast cells and basophils. In: AB Kay (ed): Allergy and *allergic diseases. Blackwell* Science, Oxford, 149–170

35   Lowman MA, Benyon RC, Church MK (1988) Characterization of neuropeptide-induced histamine release from human dispersed skin mast cells. *Br J Pharmacol* 95: 121–130

36   Meade R, Van Lovern H, Parmentier H, Iverson GM, Askenase PW (1988) The antigen-binding T cell factor PCl-F sensitizes mast cells for *in vitro* release of serotonin: Comparison with monoclonal IgE antibody. *J Immunol* 141: 2704–2713

37   Inamura N, Mekori YA, Bhattacharyya SP, Bianchine PJ, Metcalfe DD (1998) Induction and enhancement of Fc(epsilon)RI-dependent mast cell degranulation following coculture with activated T cells: dependency on ICAM-1- and leukocyte function-associated antigen (LFA)-1-mediated heterotypic aggregation. *J Immunol* 160: 4026–4033

38   Bhattacharyya SP, Drucker I, Reshef T, Kirshenbaum AS, Metcalfe DD, Mekori YA (1998) Activated T lymphocytes induce degranulation and cytokine production by human mast cells following cell-to-cell contact. *J Leukoc Biol* 63: 337–341

39   Malaviya R, Ikeda T, Ross E, Abraham SN (1996) Mast cell modulation of neutrophil

influx and bacterial clearance at sites of infection through TNF-alpha. *Nature* 381: 77–80

40 Echtenacher B, Mannel DN, Hultner L (1996) Critical protective role of mast cells in a model of acute septic peritonitis. *Nature* 381: 75–77

41 Galli SJ, Wershil BK (1996) The two faces of the mast cell. *Nature* 381: 21–22

42 Gauchat J-F, Aubry J-P, Mazzei G, Life P, Jomotte T, Elson G, J-Y Bonnefoy J-Y (1993) Human CD40-ligand: Molecular cloning, cellular distribution and regulation of expression by factors controlling IgE production. *FEBS Lett* 315: 259–266

43 Pawankar R, Okuda M, Yssel H, Okumura K, Ra CS (1997) Nasal mast cells in perennial allergic rhinitics exhibit increased expression of the FcεRI, CD40L, IL-4, and IL-13, and can induce IgE synthesis in B cells. *J Clin Invest* 99: 1492–1499

44 Liu FT (1997) Truly MASTerful cells: Mast cells command B cell IgE synthesis. *J Clin Invest* 99: 1465–1466

45 Frandji P, Tkaczyk C, Oskeritzian C, David B, Desaymard C, Mecheri S (1996) Exogenous and endogenous antigens are differentially presented by mast cells to CD4+ T lymphocytes. *Eur J Immunol* 26: 2517–2528

46 Love KS, Lakshmanan RR, Butterfield JH, Fox CC (1996) IFN-gamma-stimulated enhancement of MHC class II antigen expression by the human mast cell line HMC-1. *Cell Immunol* 170: 85–90

47 Frandji P, Tkaczyk C, Oskeritzian C, Lapeyre J, Peronet R, David B, Guillet JG, Mecheri S (1995) Presentation of soluble antigens by mast cells: upregulation by interleukin-4 and granulocyte/macrophage colony-stimulating factor and downregulation by interferon-gamma. *Cell Immunol* 163: 37–46

48 Mecheri S, David B (1997) Unraveling the mast cell dilemma: culprit or victim of its generosity? *Immunol Today* 5: 212–215

49 Wang HW, Tedla N, Lloyd AR, Wakefield D, McNeil PH (1998) Mast cell activation and migration to lymph nodes during induction of an immune response in mice. *J Clin Invest* 8: 1617–1626

50 De Pater-Huijsen FL, De Riemer MJ, Reijneke RMR, Pompen M, Lutter R, Jansen HM, Out TA (1997) Human mast cells modulate proliferation and cytokine production by CD8+ T lymphocytes. *Int Arch Allergy Immunol* 113: 287–288

51 Mihm MC, Soter NA, Dvorak HF, Austen KF (1976) The structure of normal skin and the morphology of atopic eczema. *J Invest Dermatol* 67: 305–312

52 Mitchell EB, Crow J, Williams G, Platts-Mills TAE (1986) Increase in skin mast cells following chronic house dust mite exposure. *Br J Dermatol* 114: 65–73

53 Farkas Natbony S, Phillips ME, Elias JM, Godfrey HP, Kaplan AP (1983) Histologic studies of chronic idiopathic urticaria. *J Allergy Clin Immunol* 2: 177–183

54 Garriga MM, Friedman MM, Metcalfe DD (1988) A survey of the number and distribution of mast cells in the skin of patients with mast cell disorders. *J Allergy Clin Immunol* 82: 425–432

55 Irani AM, Schwartz LB (1994) Human mast cell heterogeneity. *Allergy Proc* 15: 303–308

56   Okayama Y, Benyon RC, Rees PH, Lowman MA, Hillier K, Church MK (1992) Inhibition profiles of sodium cromoglycate and nedocromil sodium on mediator release from mast cells of human skin, lung, tonsil, adenoid and intestine. *Clin Exp Allergy* 22: 401–409

57   Langeveld-Wildschut EG, Thepen T, Bihari IC, Van Reijsen FC, De Vries IJM, Bruijnzeel PLB, Bruijnzeel-Koomen CAFM (1996) Evaluation of the atopy patch test and the cutaneous late- phase reaction as relevant models for the study of allergic inflammation in patients with atopic eczema. *J Allergy Clin Immunol* 98: 1019–1027

58   Simons FER (1998) Antihistamines. In: E Middleton, CE Reed, EF Ellis, NF Adkinson, JW Yunginger, WW Busse (eds): *Allergy. Principles and practice.* Mosby, St. Loius 612–637

59   Kubes P, Granger DN (1996) Leukocyte-endothelial cell interactions evoked by mast cells. *Cardiovasc Res* 32: 699–708

60   Walsh LJ, Trinchieri G, Waldorf HA, Whitaker D, Murphy GF (1991) Human dermal mast cells contain and release tumor necrosis factor α, which induces endothelial leukocyte adhesion molecule 1. *Proc Natl Acad Sci USA* 88: 4220–4224

61   Ackermann L, Harvima IT (1998) Mast cells of psoriatic and atopic dermatitis skin are positive for TNF-alpha and their degranulation is associated with expression of ICAM-1 in the epidermis. *Arch Dermatol Res* 290: 353–359

62   Horsmanheimo L, Harvima IT, Järvikallio A, Harvima RJ, Naukkarinen A, Horsmanheimo M (1994) Mast cells are one major source of interleukin-4 in atopic dermatitis. *Br J Dermatol* 131: 348–353

63   Thepen T, Langeveld-Wildschut EG, Bihari IC, van Wichen DF, Van Reijsen FC, Mudde GC, Bruijnzeel-Koomen CAFM (1996) Biphasic response against aeroallergen in atopic dermatitis showing a switch from an initial TH2 response to a TH1 response *in situ*: an immunocytochemical study. *J Allergy Clin Immunol* 97: 828–837

Cell-to-cell interactions

Edward F. Knol

Department of Dermatology/Allergology, G02-124, University Medical Center Utrecht, Heidelberglaan 100, NL-3584 CX Utrecht, The Netherlands

Introduction

Eczematous lesions in atopic dermatitis (AD) are characterized by a typical cellular influx in skin comprising T lymphocytes, eosinophils, macrophages, and dendritic cells (DC). Those cells are in close contact with the resident cells in skin, such as keratinocytes, mast cells, endothelial cells and fibroblasts. These typical lesions can be maintained by the (local) release of inflammatory mediators and cytokines. As is becoming more and more clear, the interactions between cells in these inflammatory loci are of importance in the outcome and modulation of this inflammation.

These cell-to-cell interactions can roughly be divided in two different mechanisms: (1) The release of inflammatory mediators and cytokines is not only important in the activation and modulation of other, neighbouring, cells, but also for the attraction of other leukocytes towards the (inflamed) skin. (2) Interactions *via* molecules expressed on the plasma membrane. This can be through interactions by co-stimulatory molecules, such as CD40/CD40L or CD28/CD86 interactions. Moreover, these interactions can occur between leukocytes and endothelial cells through cell adhesion molecules important for extravasation of leukocytes at local sites of inflammation.

This review will deal with the cell-to-cell interactions in atopic dermatitis. These interactions are important in extravasation of leukocytes towards skin. In the extravasation process adhesion molecules both on leukocytes and on endothelial cells mediate these interactions. However, they are not only involved in the extravasation of leukocytes towards affected skin, but also in the interactions between cells within the skin.

Immunohistology

Lesional skin of AD contains increased number of T lymphocytes, macrophages and dendritic cells [1, 2]. Deposits of eosinophil products are found, but only occasion-

Immunology and Drug Therapy of Allergic Skin Diseases, edited by C.A.F.M. Bruijnzeel-Koomen and E.F. Knol
© 2000 Birkhäuser Verlag Basel/Switzerland

ally intact eosinophils are found in lesional skin [3]. Although these increased cell numbers can be found both in dermis and epidermis, the most profound increase is in the dermis, where also clusters of infiltrating cells are present [1].

Already in non-involved skin of atopic dermatitis patients increased numbers of leukocytes can be demonstrated and are indicative for a so-called priming state of the "normal"-appearing skin in these patients [2].

General mechanisms regulating leukocyte extravasation

Extravasation of leukocytes to sites of inflammation is thought to consist of at least three different processes (Fig. 1) [4]. Firstly, circulating leukocytes undergo margination, whereby they move from the center to the periphery of the blood vessel and begin to bind reversibly to the endothelium, a process referred to as "rolling" (see Fig. 1; I, II). This process is mediated by the interaction of L-selectin on the leukocytes with diverse, carbohydrate-containing structures, such as glycosylation-dependent cell adhesion molecule-1 (GlyCAM-1), mucosal addressin cell adhesion molecule-1 (MadCAM-1), CD34, P-selectin or E-selectin on the endothelial cells [5]. This initial weak binding may be followed, due to activation of the leukocytes [6], by the induction of firm adhesion (see Fig. 1; II, III). Unlike the initial adhesion event, leukocyte integrins interaction with endothelial cell adhesion molecules, such as intercellular adhesion molecule-1 (ICAM-1) and vascular cell adhesion molecule-1 (VCAM-1) [7, 8], mediates this second step. Recently, it has been demonstrated that also in the initial binding of the leukocytes $\beta 1$ integrins bind to their endothelial ligands [9]. Subsequently, the leukocytes transmigrate between endothelial cells into the tissue (see Fig. 1; IV). During these events, the leukocyte integrins play a critical role. In addition, this process is controlled by platelet endothelial cell adhesion molecule-1 (PECAM-1) (CD31), as well as metabolites released from the endothelial cells and tissue-dwelling cells that form a chemotactic gradient [10–12].

Within the tissue often survival of inflammatory cells is prolonged and even cell proliferation may occur and cause large cellular infiltrates (see Fig. 1; V).

Specificity for skin-directed movement

The mechanisms described above are very general and do not explain the specific movement into the skin of subsets of leukocytes. For instance, atopic dermatitis of patients with a food allergy flares up after ingestion of a specific food to which the patient is allergic. In the affected skin an influx is noted most markedly of CD4+ T lymphocytes [13]. Ergo, specific signals/markers must be present in/on these T lymphocytes as well as in the skin, because these T lymphocytes are not directed to other tissue sites, such as lung. On the other hand in allergic contact dermatitis it is

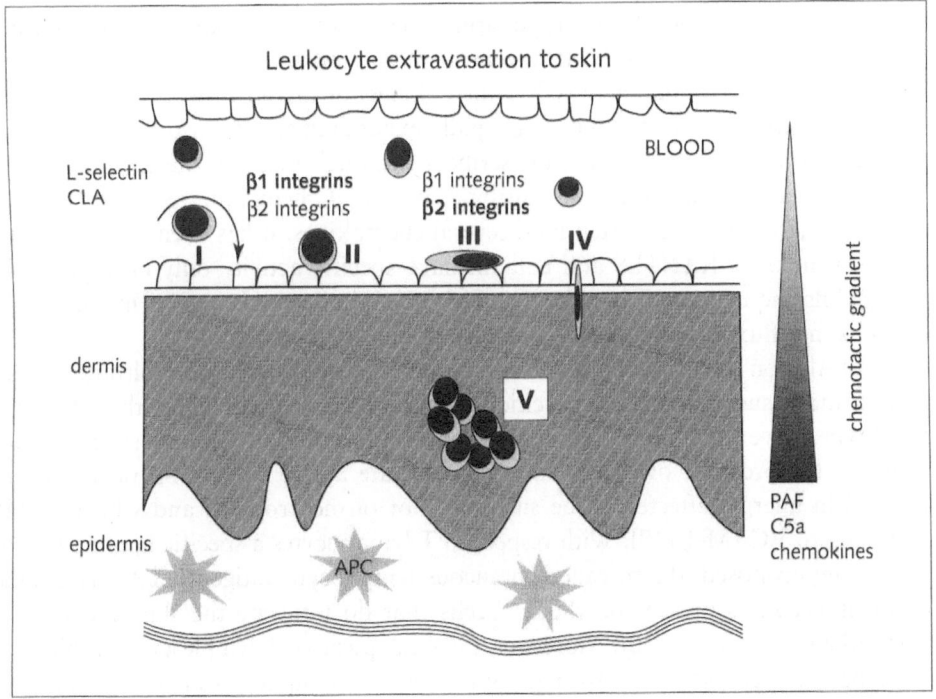

Figure 1
Leukocyte extravasation to skin. Leukocytes are present in the peripheral blood and can reversibly bind to the endothelium, a process referred to as "rolling" (I). This initial binding can be followed by the induction of more firm adhesion and flattening of the leukocytes (II and III). Subsequently, leukocyte transmigrate between endothelial cells into the tissue (IV). Within the tissue the leukocytes often demonstrate prolonged survival, interact with other cells, reside for longer periods of time and form clusters of leukocytes (V).

the cytokine released by antigen specific T cells in the skin that release a specific signal, this time leading to influx of lymphocytes and neutrophils. Several more-or-less specific mechanisms exists that can explain, in part, these specific movements. Directed movement can be obtained by activity of (a) chemoattractants and/or expression of (b) adhesion molecules. This will be addressed below.

(a) First, specific leukocytes have on their plasma membrane receptors for locally released attractants, such as platelet-activating factor (PAF), leukotriene B_4 and C5a [14]. The best-described family of attractants is the so-called chemokine family of cytokines. Chemokines are structurally related, cell-derived, relatively cationic, peptides of approximately 8–10 kDa molecular weight that can be produced by endothelial cells, fibroblasts and keratinocytes. These chemokines are divided into two subfamilies based on the arrangement of the first two cysteins, which are either separated by one amino acid (CXC chemokines) or are adjacent (CC chemokines) [15]. The group of chemokines is a rapidly expanding group. At present more than 40 different members have been described. Although many of the chemokines can have overlapping activities and can bind to several chemokine receptors, and one chemokine receptor can often bind several chemokines, it has been shown that the CC chemokines RANTES and eotaxin have specific activity only on eosinophils, basophils and memory T lymphocytes [15–17]. Injection of RANTES in human skin results in influx of lymphocytes and eosinophils [18].

(b) Also on the basis of specific adhesion molecules present on infiltrating leukocytes and tissue endothelium specificity for the tissue, as well as for the infiltrating cell types have been hypothesized. The most important, but not very specific difference is the presence of the β1 integrin very late antigen-4 (VLA4) on leukocytes found in allergic affected tissue sites, but not on neutrophils, and which enables binding to VCAM-1 [19]. With respect to T lymphocytes a specific homing marker has been proposed, the so-called cutaneous lymphocyte antigen (CLA). This adhesion molecule is present on those T cells that do infiltrate the skin and it binds endothelial E-selectin [20]. Moreover, the CLA-positive T lymphocytes in blood of AD patients represent an activated population of T lymphocytes [21]. However, the supposed specificity of the CLA for T lymphocytes' skin homing is under discussion. For instance, it has been demonstrated that patients with a defect in fucose metabolism resulting in the leukocyte adhesion deficiency syndrome type II (LAD-II) do not express the CLA adhesion molecule. However, these LAD-II patients exhibit relatively normal skin homing of T lymphocytes [22]. Moreover, it has also become clear that although the CLA is present only on a subset of the T lymphocytes, it is expressed in high amounts on other leukocytes, such as neutrophils [23], cells that are not present in the inflammatory lesions of atopic dermatitis [24].

Adhesion molecule expression in atopic dermatitis

The expression of adhesion molecules on endothelium in skin of atopic dermatitis patients has been demonstrated to be increased. Already in non-lesional skin of AD patients increased expression of E-selectin (CD62E), P-selectin (CD62P), VCAM-1 and ICAM-1 [25, 26] has been described. After application of allergens onto the skin by patch-testing a further increase in these endothelial cell adhesion molecules was found [25, 27]. The highest expression of these adhesion molecules was found on endothelial cells in skin of eczematous skin [25].

Although several pro-inflammatory cytokines and mediators are known to increase adhesion molecule expression on endothelium [4] only mast cell products such as TNFα have been associated with increased adhesion molecules expression in AD patients [28, 29]. Remarkably, in contrast to umbilical cord endothelial cells, skin microvascular endothelial cells do not demonstrate IL-4 receptors ruling out an effect of the Th2 cytokine IL-4 on adhesion molecule expression in skin [25, 30].

Role of tissue-dwelling cells in the cell-to-cell interactions in skin

The synthesis and expression of both adhesion molecules and chemotactic molecules in skin are of importance for the development of a leukocyte infiltration. In the skin several tissue-dwelling cells are involved in these two mechanisms.

Endothelial cells in skin are not only involved *via* the increased expression of adhesion molecules on their apical side as stated earlier, they are also important sources of cytokines such as IL-1, IL-6, GM-CSF, IL-8, MCP-1, GRO, RANTES and IP-10 [31–33].

Keratinocytes can synthesize and release several chemotactic molecules, such as interleukin 8 (IL-8) [34], monocyte chemotactic protein-1 (MCP-1) [35] and inter-feron-γ-inducible protein (IP-10) [36]. Moreover, keratinocytes release both tumor-necrosis factor α (TNFα) and IL-1α, potent inducers of adhesion molecules on endothelial cells and granulocyte-macrophage colony-stimulating factor (GM-CSF) synthesis by endothelial cells [37]. Upon activation, most markedly by IFNγ, a marked increase in the expression of ICAM-1 and HLA-DR has been described on keratinocytes [37]. Keratinocytes obtained from uninvolved AD skin spontaneously release TNFα, GM-CSF and IL-1α [37].

A potent source of cytokines is the fibroblasts. Dermal fibroblasts have been demonstrated to produce GM-CSF, RANTES, IL-8, eotaxin [38], macrophage inflammatory protein-1α (MIP-1α), MCP-3, and MCP-4 [38–42].

Antigen-presenting cells (APC) represented by Langerhans cells in the epidermis and dendritic cells in the dermis are of importance for the local activation of T lym-phocytes. *Via* the presentation of processed antigen in the context of MHC mole-cules together with costimulatory molecules APC's activate the T lymphocytes for release of cytokines and proliferation [43]. Although APC are poor local releasers of inflammatory mediators and cytokines, they are known to skew the phenotype of Th cells by the release of IL-12 towards a Th1 response [44]. The Langerhans cells in skin of AD patients exclusively express IgE on their cell membrane bound to the high affinity receptor for IgE [45, 46]. This membrane-bound IgE improves the anti-gen-presenting capacity by the Langerhans cells to T cells [47].

The perivascular location and their content of pre-formed TNFα and histamine, makes mast cells important players in the activation of endothelial cells, as well as the attraction of inflammatory leukocytes. Apart from histamine and TNFα, mast cells can synthesize and release many proinflammatory mediators, cytokines and chemokines, such as PAF, tryptase, LTB$_4$, IL-5, IL-6, IL-13, IL-8, possibly IL-4, MCP-1, MIP-1α and lymphotactin [48–51].

Recently, it has become clear that also *via* nerve endings in the skin the nervous system can regulate cutaneous inflammation [52]. Calcitonin gene-related peptide (CGRP) release from nerve endings can directly activate Langerhans cells [53]. Release of substance P from nerve fibers results in the induction of VCAM-1 expression on microvascular endothelial cells and in the release of TNFα from mast cells, as well as keratinocytes [54–56].

Local cell-to-cell interactions in the maintenance of eczematous lesions

Although much has been learned about the mechanisms involved in the induction of an eczematous reactions, less is clear about the mechanisms that are responsible for the maintenance of the inflammatory reaction. As is stated before, within the inflammatory reaction site close contact exists between resident and infiltrating cells. On the membrane of these cells specialized receptors are present, so-called co-stimulatory molecules, that modulate their activity [57]. The possible role of co-stimulation locally in the skin will be addressed below.

The role of co-stimulatory molecules in skin inflammation

Activation of several cell types, most notably T lymphocytes, requires two signals to obtain full activation. One of these signals is provided by the binding of CD80 (B7-1) or CD86 (B7-2) to CD28 or CTLA-4 on T cells [58]. Apart from T lymphocytes, ligand-binding to CD28 also activates others cell types, such as mast cells [59]. Expression of CD80 and CD86 can therefore be an important mechanism in the regulation of the locally present cells, most markedly T cells. In skin, CD80 and CD86 expression is demonstrated on Langerhans cells and its expression is increased in atopic dermatitis [60]. CD80 and CD86 expression on Langerhans cells can be modulated by ultraviolet B radiation and IFNγ addition [61, 62]. The importance of CD86 was shown by the almost complete inhibition by CD86 antibodies of allergen-induced T cell proliferation stimulated with allergen in the presence of epidermal cells [60]. Also in animal models it was shown that CD86 antibodies were potent in the abolishment of irritant-induced contact hypersensitivity reactions [63, 64].

Another important co-stimulatory pathway is the CD40-CD40L pathway. CD40L is predominantly expressed on T lymphocytes, whereas with regard to skin, CD40 expression is described on endothelial cells [65] and keratinocytes [66]. The expression of CD40 on endothelium is increased in inflamed skin [65], whereas IFNγ addition increases CD40 expression on keratinocytes [66]. Activation of these cell types by cross-linking of their CD40, binding of its soluble ligand gp39, or contact with CD40L-expressing T lymphocytes increases adhesion molecule expression

[65] and the release of IL-6 and IL-8 [66, 67]. In the interaction of T lymphocytes with keratinocytes CD40 provides an important costimulatory signal for PHA-induced T lymphocyte proliferation [67].

Apart from co-stimulatory molecules, leukocytes can also be activated or primed *via* the interaction through their β1 and β2 integrins [68]. The adhesion molecules ICAM-1, ICAM-2 and ICAM-3 can provide a costimulatory signal *via* binding to the β2 integrin LFA-1 on T lymphocytes [69]. ICAM-1 is regularly expressed or inducible on all major cutaneous cell populations including Langerhans cells, keratinocytes, endothelial cells and dermal fibroblasts [26]. ICAM-1 expression has been found to be increased in the skin under inflammatory conditions and plays an important role in the activation of T cells [70–72].

Apart from binding to ICAM-1 and VCAM-1, it is likely that several of the extracellular matrix proteins (ECM) present in the skin will bind to the β1 and β2 integrins on the membranes of cells present [14]. Although the presence and role of extracellular matrix proteins have until now only been investigated in the context of skin repair [16], it is tempting to speculate on the role of these ECM released by, for instance, endothelial cells. Local release of the ECM might modulate the inflammatory response in AD.

Concluding remarks

In the cellular events occurring in allergic skin disease it is clear that cell-to-cell interactions are important. These interactions are crucial for the skin-directed movement of leukocytes. However, also within the skin it is clear that many cell-to-cell interactions are active that determine the final outcome of the local skin response that can finally evolve in the eczemateous lesional skin. A better insight in these interactions is crucial for the understanding of the pathophysiological mechanisms determining atopic skin disease. Interfering in these cell-to-cell interactions is a challenge for the near future which might result in the development of new therapeutics.

References

1   Leung DYM, Bhan AK, Schneeberger EE, Geha RS (1983) Characterization of the mononuclear cell infiltrate in atopic dermatitis monoclonal antibodies. *J Allergy Clin Immunol* 71: 47–56

2   Thepen T, Langeveld-Wildschut EG, Bihari IC, van Wichen DF, Van Reijsen FC, Mudde GC, Bruijnzeel-Koomen CAFM (1996) Biphasic response against aeroallergen in atopic dermatitis showing a switch from an initial TH2 response to a TH1 response *in situ*: an immunocytochemical study. *J Allergy Clin Immunol* 97: 828–837

3   Leiferman KM (1994) Eosinophils in atopic dermatitis. *J Allergy Clin Immunol* 94 (Suppl): 1310–1317

4   Butcher EC (1991) Leukocyte-endothelial cell recognition. Three (or more) steps to specificity, and diversity. *Cell* 67: 1033–1036

5   Varki A (1994) Selectin ligands. *Proc Natl Acad Sci USA* 91: 7390–7397

6   Kuijpers TW, Hakkert BC, Hoogerwerf M, Leeuwenberg JFM, Roos D (1991) Role of endothelial leukocyte adhesion molecule-1, platelet-activating factor in neutrophil adherence to IL-1-prestimulated endothelial cells. Endothelial leukocyte adhesion molecule-1-mediated CD18 activation. *J Immunol* 147: 1369–1376

7   Rice GE, Munro JM, Bevilacqua MP (1990) Inducible cell adhesion molecule 110 (INCAM-110) is an endothelial receptor for lymphocytes. A CD11/CD18-independent adhesion mechanism. *J Exp Med* 171: 1369–1374

8   Lawrence MB, Springer TA (1991) Leukocytes roll on a selectin at physiological flow rates: distinction from, and prerequisite for adhesion through integrins. *Cell* 65: 859–865

9   Chen C, Mobley JL, Dwir O, Shimron F, Grabovsky V, Lobb RR, Shimizu Y, Alon R (1999) High affinity very late antigen-4 subsets expressed on T cells are mandatory for spontaneous adhesion strenghtening but not for rolling on VCAM-1 in shear flow. *J Immunol* 162: 1084–1095

10   Luscinskas FW, Cubulsky MI, Kiely J-M, Peckins CS, Davis VM, Gimbrone MA (1991) Cytokine-activated human endothelial monolayers support enhanced neutrophil transmigration via a mechanism involving both endothelial-leukocyte adhesion molecule-1, and intracellular adhesion molecule-1. *J Immunol* 146: 1617–1623

11   Muller WA, Weigl SA, Deng X, Phillips DM (1993) PECAM-1 is required for transmigration of leukocytes. *J Exp Med* 178: 449–460

12   Springer TA (1994) Traffic signals for lymphocyte recirculation, and leukocyte emigration: the multistep paradigm. *Cell* 76: 301–314

13   Sampson HA (1995) Mechanisms of adverse reactions to food. The skin. *Allergy* 20S: 46–51

14   Kuijpers TW, Roos D (1993) Leukocyte extravasation: mechanisms, and consequences. *Behring Inst Mitt* 107–137

15   Dahinden CA (1997) Chemokines. In: AB Kay (ed): *Allergy, and allergic diseases.* Blackwell Sciences, Oxford, 365–379

16   Juhasz I, Murphy GF, Yan HC, Herlyn M, Albelda SM (1993) Regulation of extracellular matrix proteins, and integrin cell substratum adhesion receptors on epithelium during cutaneous human wound healing *in vivo. Am J Pathol* 5: 1458–1469

17   Sallusto F, Mackay CR, Lanzavecchia A (1997) Selective expression of the eotaxin receptor CCR3 by human T helper 2 cells. *Science* 277: 2005–2007

18   Beck LA, Dalke S, Leiferman KM, Bickel CA, Hamilton R, Rosen H, Bochner BS, Schleimer RP (1997) Cutaneous injection of RANTES causes eosinophil recruitment – Comparison of nonallergic, and allergic human subjects. *J Immunol* 159: 2962–2972

19   Bochner BS, Luscinskas FW, Gimbrone MA Jr, Newman W, Sterbinsky SA, Derse Antho-

ny CP, Klunk D, Schleimer RP (1991) Adhesion of human basophils, eosinophils, and neutrophils to interleukin 1-activated human vascular endothelial cells: contributions of endothelial cell adhesion molecules. *J Exp Med* 173: 1553–1557

20   Picker LJ (1992) Mechanisms of lymphocyte homing. *Curr Opin Immunol* 4: 227–286

21   Akdis M, Akdis CA, Weigl L, Disch R, Blaser K (1997) Skin-homing, CLA⁺ memory T cells are activated in atopic dermatitis, and regulate IgE by an IL-13-dominated cytokine pattern: IgG4 counter-regulation by CLA-memory T cells. *J Immunol* 159: 4611–4619

22   Kuijpers TW, Etzioni A, Pollack S, Pals ST (1997) Antigen-specific immune responsiveness, and lymphocyte recruitment in leukocyte adhesion deficiency type II. *Int Immunol* 9: 607–613

23   Picker LJ, Michie SA, Rott LS, Butcher EC (1990) A unique phenotype of skin-associated lymphocytes in humans. Preferential expression of the HECA-452 epitope by benign, and malignant T cells at cutaneous sites. *Am J Pathol* 136: 1053–1068

24   Langeveld-Wildschut EG, Thepen T, Bihari IC, Van Reijsen FC, De Vries IJM, Bruijnzeel PLB, Bruijnzeel-Koomen CAFM (1996) Evaluation of the atopy patch test, and the cutaneous late-phase reaction as relevant models for the study of allergic inflammation in patients with atopic eczema. *J Allergy Clin Immunol* 98: 1019–1027

25   De Vries IJ, Langeveld-Wildschut EG, Van Reijsen FC, Dubois GR, Van den Hoek JA, Bihari IC, van Wichen D, Weger RA, Knol EF, Thepen T, Bruijnzeel-Koomen CAFM (1998) Adhesion molecule expression on skin endothelia in atopic dermatitis: effects of TNF-alpha, and IL-4. *J Allergy Clin Immunol* 102: 461–468

26   Jung K, Linse F, Heller R, Moths C, Goebel R, Neumann C (1996) Adhesion molecules in atopic dermatitis: VCAM-1, and ICAM-1 expression is increased in healthy-appearing skin. *Allergy* 51: 452–460

27   Jung K, Linse F, Pals ST, Heller R, Moths C, Neumann C (1997) Adhesion molecules in atopic dermatitis: patch tests elicited by house dust mite. *Contact Dermatitis* 37: 163–172

28   Meng H, Tonnesen MG, Marchese MJ, Clark RA, Bahou WF, Gruber BL (1995) Mast cells are potent regulators of endothelial cell adhesion molecule ICAM-1, and VCAM-1 expression. *J Cell Physiol* 165: 40–53

29   Ackermann L, Harvima IT (1998) Mast cells of psoriatic, and atopic dermatitis skin are positive for TNF-alpha, and their degranulation is associated with expression of ICAM-1 in the epidermis. *Arch Dermatol Res* 290: 353–359

30   Schleimer RP, Sterbinsky SA, Kaiser J, Bickel CA, Klunk DA, Tomioka K, Newman W, Luscinskas FW, Gimbrone MA, McIntyre BW et al (1992) IL-4 induces adherence of human eosinophils, and basophils but not neutrophils to endothelium association with expression of VCAM-1. *J Immunol* 148: 1086–1092

31   Swerlick RA, Lawley TJ (1993) Role of microvascular endothelial cells in inflammation. *J Invest Dermatol* 100: 111S–115S

32   Santamaria Babi LF, Moser B, Perez Soler MT, Moser R, Loetscher P, Villiger B, Blaser K, Hauser C (1996) The interleukin-8 receptor B, and CXC chemokines can mediate

transendothelial migration of human skin homing T cells. *Eur J Immunol* 26: 2056–2061

33 Goebeler M, Yoshimura T, Toksoy A, Ritter U, Brocker EB, Gillitzer R (1997) The chemokine repertoire of human dermal microvascular endothelial cells, and its regulation by inflammatory cytokines. *J Invest Dermatol* 108: 445–451

34 Boorsma DM, de Haan P, Willemze R, Stoof TJ (1994) Human growth factor (huGRO), interleukin-8 (IL-8), and interferon-gamma-inducible protein (gamma-IP-10) gene expression in cultured normal human keratinocytes. *Arch Dermatol Res* 286: 471–475

35 Yu X, Barnhill RL, Graves DT (1994) Expression of monocyte chemoattractant protein-1 in delayed type hypersensitivity reactions in the skin. *Lab Invest* 71: 226–235

36 Kaplan G, Luster AD, Hancock G, Cohn ZA (1987) The expression of a gamma interferon-induced protein (IP-10) in delayed immune responses in human skin. *J Exp Med* 166: 1098–1108

37 Pastore S, Corinti S, La Placa M, Didona B, Girolomoni G (1998) Interferon-gamma promotes exaggerated cytokine production in keratinocytes cultured from patients with atopic dermatitis. *J Allergy Clin Immunol* 101: 538–544

38 Noso N, Bartels J, Mallet AI, Mochizuki M, Christophers E, Schroder JM (1998) Delayed production of biologically active O-glycosylated forms of human eotaxin by tumor-necrosis-factor-alpha-stimulated dermal fibroblasts. *Eur J Biochem* 253: 114–122

39 Noso N, Sticherling M, Bartels J, Mallet AI, Christophers E, Schroder JM (1996) Identification of an N-terminally truncated form of the chemokine RANTES, and granulocyte-macrophage colony-stimulating factor as major eosinophil attractants released by cytokine-stimulated dermal fibroblasts. *J Immunol* 156: 1946–1953

40 Sticherling M, Hetzel F, Schroder JM, Christophers E (1993) Time-, and stimulus-dependent secretion of NAP-1/IL-8 by human fibroblasts, and endothelial cells. *J Invest Dermatol* 4: 573–576

41 Bug G, Aman MJ, Tretter T, Huber C, Peschel C (1998) Induction of macrophage-inflammatory protein 1alpha (MIP-1alpha) by interferon-alpha. *Exp Hematol* 26: 117–123

42 Hein H, Schluter C, Kulke R, Christophers E, Schroder JM, Bartels J (1999) Genomic organization, sequence analysis, and transcriptional regulation of the human MCP-4 chemokine gene (SCYQ13) in dermal fibroblasts: a comparison to other eosinophilic beta-chemokines. *Biochem Biophys Res Commun* 255: 470–476

43 Grewe M, Bruijnzeel-Koomen CAFM, Schöpf E, Thepen T, Langeveld-Wildschut AG, Ruzicka T, Krutman J (1998) A role for Th1, and Th2 cells in the immunopathogenesis of atopic dermatitis. *Immunol Today* 8: 359–361

44 Hart DNJ (1997) Dendritic cells: unique leukocyte populations which control the primary immune response. *Blood* 9: 3245–3287

45 Bruijnzeel-Koomen CAFM, van Wichen DF, Toonstra J, Berrens L, Bruijnzeel PL (1986) The presence of IgE molecules on epidermal Langerhans cells in patients with atopic dermatitis. *Arch Dermatol Res* 278: 199–205

46   Bieber T, de la Salle H, Wollenberg A, Hakimi J, Chizzonite R, Ring J, Hanau D, de la Salle C (1992) Human epidermal Langerhans cells express the high affinity receptor for immunoglobulin E (Fc epsilon RI). *J Exp Med* 175: 1285–1290

47   Mudde GC, Van Reijsen FC, Boland GJ, de Gast GC, Bruijnzeel PLB, Bruijnzeel-Koomen CAFM (1990) Allergen presentation by epidermal Langerhans' cells from patients with atopic dermatitis is mediated by IgE. *Immunology* 69: 335–341

48   Church MK, Holgate ST, Shute JK, Walls AF, Sampson AP (1998) Mast cell-derived mediators. In: E Middleton, EF Ellis, JW Yunginger, CE Reed, NF Adkinson, WW Busse (eds): *Allergy. Principles and practice*. Mosby Year Book, St.Louis, 146–182

49   Baghestanian M, Hofbauer R, Kiener HP, Bankl HC, Wimazal F, Willheim M, Scheiner O, Fureder W, Muller MR, Bevec D, Lechner K, Valent P (1997) The c-kit ligand stem cell factor, and anti-IgE promote expression of monocyte chemoattractant protein-1 in human lung mast cells. *Blood* 90: 4438–4449

50   Yano K, Yamaguchi M, De Mora F, Lantz CS, Butterfield JH, Costa JJ, Galli SJ (1997) Production of macrophage inflammatory protein-1a by human mast cells: Increased anti-IgE-dependent secretion after IgE-dependent enhancement of mast cell IgE-binding ability. *Lab Invest* 77: 185–193

51   Rumsaeng V, Vliagoftis H, Oh CK, Metcalfe DD (1997) Lymphotactin gene expression in mast cells following Fc0 receptor I aggregation – Modulation by TGF-β, IL-4, dex-amethasone, and cyclosporin A. *J Immunol* 158: 1353–1360

52   Eedy DJ (1993) Neuropeptides in skin. *Br J Dermatol* 128: 597–605

53   Hosoi J, Murphy GF, Egan CL, Lerner EA, Grabbe S, Asahina A, Granstein RD (1993) Regulation of Langerhans cell function by nerves containing calcitonin gene-related pep-tide. *Nature* 363: 159–163

54   Quinlan KL, Song IS, Naik SM, Letran EL, Olerud JE, Bunnett NW, Armstrong CA, Caughman SW, Ansel JC (1999) VCAM-1 expression on human dermal microvascular endothelial cells is directly, and specifically up-regulated by substance P. *J Immunol* 3: 1656–1661

55   Ansel JC, Brown JR, Payan DG, Brown MA (1993) Substance P selectively activates TNF-α gene expression in murine mast cells. *J Immunol* 150: 4478–4485

56   Viac J, Gueniche A, Doutremepuich JD, Reichert U, Claudy A, Schmitt D (1996) Sub-stance P, and keratinocyte activation markers: an *in vitro* approach. *Arch Dermatol Res* 288: 85–90

57   Gause WC, Halvorson MJ, Lu P, Greenwald R, Linsley P, Urban JF, Finkelman FD (1997) The function of costimulatory molecules, and the development of IL-4-produc-ing T cells. *Immunol Today* 18: 115–120

58   Lenschow DJ, Walunas TL, Bluestone JA (1996) CD28/B7 system of T cell costimula-tion. *Annu Rev Immunol* 14: 233–258

59   Tashiro M, Kawakami Y, Abe R, Han W, Hata D, Sugie K, Yao L, Kawakami T (1997) Increased secretion of TNF-α by costimulation of mast cells via CD28, and FcεRI. *J Immunol* 158: 2382–2389

60   Ohki O, Yokozeki H, Katayama I, Umeda T, Azuma M, Okumura K, Nishioka K (1997)

Functional CD86 (B7-2/B70) is predominantly expressed on Langerhans cells in atopic dermatitis. *Br J Dermatol* 136: 838–845

61 Weiss JM, Renkl AC, Denfeld RW, de Roche R, Spitzlei M, Schopf E, Simon JC (1995) Low-dose UVB radiation perturbs the functional expression of B7.1, and B7.2 co-stimulatory molecules on human Langerhans cells. *Eur J Immunol* 25: 2858–2862

62 Yokozeki H, Katayama I, Ohki O, Arimura M, Takayama K, Matsunaga T, Satoh T, Umeda T, Azuma M, Okumura K, Nishioka K (1997) Interferon-gamma differentially regulates CD80 (B7-1), and CD86 (B7-2/B70) expression on human Langerhans cells. *Br J Dermatol* 136: 831–837

63 Nuriya S, Yagita H, Okumura K, Azuma M (1996) The differential role of CD86, and CD80 co-stimulatory molecules in the induction, and the effector phases of contact hypersensitivity. *Int Immunol* 8: 917–926

64 Katayama I, Matsunaga T, Yokozeki H, Nishioka K (1997) Blockade of costimulatory molecules B7-1 (CD80), and B7-2 (CD86) down-regulates induction of contact sensitivity by haptenated epidermal cells. *Br J Dermatol* 136: 846–852

65 Hollenbaugh D, Mischel Petty N, Edwards CP, Simon JC, Denfeld RW, Kiener PA, Aruffo A (1995) Expression of functional CD40 by vascular endothelial cells. *J Exp Med* 182: 33–40

66 Denfeld RW, Hollenbaugh D, Fehrenbach A, Weiss JM, von Leoprechting A, Mai B, Voith U, Schopf E, Aruffo A, Simon JC (1996) CD40 is functionally expressed on human keratinocytes. *Eur J Immunol* 26: 2329–2334

67 Gaspari AA, Sempowski GD, Chess P, Gish J, Phipps RP (1996) Human epidermal keratinocytes are induced to secrete interleukin-6, and co-stimulate T lymphocyte proliferation by a CD40-dependent mechanism. *Eur J Immunol* 26: 1371–1377

68 Clark EA, Brugge JS (1995) Integrins, and signal transduction pathways: the road taken. *Science* 268: 233–239

69 Zambruno G, Cossarizza A, Zacchi V, Ottani D, Luppi AM, Giannetti A, Girolomoni G (1995) Functional intercellular adhesion molecule-3 is expressed by freshly isolated epidermal Langerhans cells, and is not regulated during culture. *J Invest Dermatol* 105: 215–219

70 Griffiths CE, Railan D, Gallatin WM, Cooper KD (1995) The ICAM-3/LFA-1 interaction is critical for epidermal Langerhans cell alloantigen presentation to CD4+ T cells. *Br J Dermatol* 133: 823–829

71 van Pelt JP, Kuijpers SH, van de Kerkhof PC, de Jong EM (1998) The CD11b/CD18-integrin in the pathogenesis of psoriasis. *J Dermatol Sci* 16: 135–143

72 Griffiths CEM, Voorhees JJ, Nickoloff BJ (1989) Characterization of intercellular adhesion moelcule-1, and HLA-DR expression in normal, and inflamed skin: modulation by recombinant gamma interferon, and tumor necrosis factor. *J Am Acad Dermatol* 20: 617–629

Atopic dermatitis

Carla A.F.M. Bruijnzeel-Koomen

Department of Dermatology, University Hospital Utrecht, Heidelberglaan 100, NL-3584 CX Utrecht, The Netherlands

Atopic dermatitis (AD) is part of the atopy syndrome. It is the most common allergic skin disease. In this chapter the clinical characteristics and pathogenesis will be discussed. Therapy will be discussed in a separate chapter.

Definition

Thus far, a distinctive diagnostic indicator for AD has not been described. The diagnosis relies on a cluster of clinical and serological features. In 1980, for the first time, diagnostic criteria were established to create a greater homogeneity in the diagnosis of AD [1]. For the diagnosis of AD, the presence of at least three of the basic features mentioned in Table 1 is required. In addition to having at least three of the basic features, a patient should manifest at least three of the minor features mentioned in Table 2.

Genetics

AD is together with asthma and allergic rhinitis part of the atopy syndrome. It is a complex multifactorial disease sharing many pathogenetic features with allergic asthma and rhinitis. Twin studies have identified a substantial genetic contribution to the development of AD. Monozygotic twins run a risk of 0.86 of having AD if the twin partner has the disease, in contrast to a disease risk of 0.21 run by dizygotic twins, which does not differ from the frequency seen in ordinary siblings. In spite of this genetic influence the mode of transmission does not conform to a simple Mendelian pattern of inheritance [2].

Atopic diseases tend to run true to type within each family; in some families most of the affected members will have eczema, while in others asthma or hayfever will predominate [3]. Some studies confirm that atopic diseases, whether respiratory or eczematous are inherited more often from the mother than from the father, and this

Immunology and Drug Therapy of Allergic Skin Diseases, edited by C.A.F.M. Bruijnzeel-Koomen and E.F. Knol
© 2000 Birkhäuser Verlag Basel/Switzerland

Table 1 - Major features of atopic dermatitis

- facial and extensor involvement in infants and children, flexural lichenification or linearity in adults
- chronic or chronically relapsing dermatitis
- personal or family history of atopy (asthma, allergic rhinitis, atopic dermatitis)

Table 2 - Minor features of atopic dermatitis

- anterior neck folds
- anterior subcapsular cataracts
- cheilitis
- course influenced by environmental/emotional factors
- Dennie-Morgan infraorbital fold
- early age of onset
- elevated serum IgE
- facial pallor/facial erythema
- food intolerance
- ichthyosis/palmar hyperlinearity/keratosis pilaris
- immediate skin test reactivity
- impaired cell mediated immunity
- intolerance to wool and lipid solvents
- itch when sweating
- keratoconus
- nipple eczema
- orbital darkening
- perifollicular darkening
- pityriasis alba
- recurrent conjunctivitis
- tendency towards cutaneous infection
- tendency towards non-specific hand or foot dermatitis
- white dermographism/delayed blanching phenomenon
- xerosis

of course fits into the finding that the atopy locus lies on chromosome 11q13 [4]. The observation that the gene for the beta-subunit of the high-affinity Fc-receptor for immunoglobulin E (IgE) (which is present on mast cells, basophils, monocytes and dendritic cells like Langerhans cells) lies on chromosome 11q13 and is closely linked to the gene for atopy suggests a genetic bridge between atopic diseases with a predominant immediate type of allergic reactions (asthma and hayfever) and atopic diseases with a predominant delayed type or cellular reaction type like AD [5]. However, the literature provides us with conflicting data concerning the linkage between atopy and the high affinity receptor gene at 11q13 in AD families [6, 7]. Recently polymorphisms within the Fc epsilonR1-beta gene were reported to be strongly associated with AD [8]. These results should be connected with the observation that dendritic cells lack the gene for the beta subunit of the high-affinity receptor for IgE [9, 10].

Epidemiology

The lack of a distinct diagnostic indicator has interfered with full epidemiological case-finding. Assessments must rely on gross physical features, which may be totally absent during remissions. In spite of this limitation studies have shown AD to be very common.

The prevalence of AD has increased in the last three decades. A questionnaire survey using a scoring system showed a cumulative prevalence of AD of 15.6% [11]. The reasons for the increase in prevalence of AD are not known.

Clinical features

The clinical signs and symptoms of AD differ according to the period of life in which the onset occurs. The very early infantile phase starts around 3 months of age. There is also a childhood, an adolescent and an adult phase.

In the infantile phase, the face, the extensor aspect of the extremities and the trunk are involved. The early infantile phase is characterized by erythema, desquamation and, in severe cases, exudation. In the late infantile phase impetiginous, vesicular, eczematous and sometimes even lichen-like lesions become more prominent. The intense itch is reflected by the presence of haemorrhagic crusts. In the early phase onset other areas are usually involved, such as the perioral area, the lips (upper lip), the periocular area and the flexural regions of the limbs, where dry and lichenified lesions are present.

Dry skin (xerosis) is often seen in patients with AD. The clinical condition is characterized by a rough, faintly scaling, non-inflamed skin surface. The dry skin condition worsens in a cold environment with reduced humidity especially in winter and after bathing. Dry skin may reflect mild eczematous changes, a manifestation of ichthyosis vulgaris or a combination of these. Sebum secretion is reduced and the composition of sebum is changed, including increased cholesterol, decreased squalene and unsaturated fatty acids, which leads to increased epidermal water loss.

For several decades, it has been noticed that, while normal skin when rubbed becomes red, atopic skin tends after a few seconds to become white (white dermographism). Intradermal injections of adrenaline in patients with AD induces a reddening, which is slowly replaced by pallor, which may be quite intense and lasts 15–30 min. The difficulty in AD is to decide whether the pallor is due to edema or to vasoconstriction. The response to cholinergic agents is hyperreactive, analogous to the bronchoconstrictor reaction after inhalation of cholinergic agents.

Variants of AD may occur alone, together or in combination with more typical forms as described above (eczematoid or lichenified skin lesions). If not, they may hinder the making of the diagnosis of AD. In that case the family or personal atopy history of the patient is important. Variants include:

- follicular or patchy pityriasiform lichenoid eczema; this is a dry, only slightly irritating, follicular form, occurring in childhood. These scaly non-hyperkeratotic, skin-coloured papules are present on the trunk, the nape of the neck and the extensor aspects of the knees, in combination with a generally dry skin.
- exfoliating cheilitis with perleche;
- retroauricular intertrigo or earlobe rhagades;
- lower eyelid eczema with lichenification;
- nipple eczema;
- tylotic, rhagadiform, fingerpad eczema;
- juvenile plantar dermatosis or atopic winter feet; this disease is a painful variant of AD, frequently occurring in children. It is localized on the anterior part of the foot and is characterized by erythema, hyperkeratosis and raghades. It also occurs, however, in the summer months and is often misdiagnosed as a mycotic infection. Complete recovery may be expected in puberty in the majority of patients.

Course and prognosis

Rajka [12] calculated that 60% of patients with AD had onset of their disease in the first year of life and 85% before the age of 5 years. The peak of onset was calculated to be before 3 months of life.

The dermatitis tends to be more constantly present in the younger age groups and symptom-free periods appear more frequently with maturity. About 40% of patients show clearing of the disease; patients with the most severe disease are twice as likely to have persistent disease.

The clinical pattern suggests that there are several types of AD with different genetic backgrounds: one type of dermatitis with predominantly early onset of AD and one type with similar skin manifestations, but not associated with allergy and having a predominantly later onset.

Pathology and immunopathology

Thus far, no distinctive histopathological marker has been described. Acute lesions show slight hyperkeratosis and parakeratosis, marked intercellular edema with variable spongiotic vesicle formation.

The number of Langerhans cells is not increased. The phenotype of these cells is changed. They express CD1b (in addition to CD1a) and CD36.

Another conspicuous change in the phenotype of Langerhans cells is the binding of easily detectable IgE molecules, particularly in lesional skin [13]. The IgE molecules are bound by an Fc receptor 1, but missing the β-chain which is expressed by the Fc-epsilon receptor 1 in mast cells and basophils. Specific IgE molecules bound by the Fc-epsilon receptor on Langerhans cells are involved in allergen presentation to T cells [14].

The cellular infiltrate in acute and chronic AD lesions consists of T lymphocytes. The ratio of CD4 cells to CD8 cells is increased and may even be higher than the one in peripheral blood. This suggests a selective infiltration of CD4 cells in the dermal infiltrate.

This subset of T lymphocytes appears to have undergone intralesional activation as evidenced by their expression of IL-2 receptors and HLA-DR molecules. A substantial percentage of these T cells express high levels of the cutaneous lymphocyte associated antigen (CLA), a carbohydrate ligand for the endothelial leucocyte adhesion molecule 1 (ELAM-1) which appears to function as a skin lymphocyte homing receptor. The cytokine expression of the T cells present in lesional AD skin is consistent with the T-helper 1 (Th1), Th2 and Th0 types of T cells. Using IL-4 and interferon γ (IFNγ) production as a discriminator between Th types, it was found, by means of immunohistochemical staining that the majority of the CD4 T cells produce both IL-4 and IFNγ (being Th0 cells) and a minority produce either IL-4 (Th2) or IFNγ (Th1 cells) [15]. Using polymerase chain reaction procedure it was demonstrated that acute lesions are dominated by mRNA for IL-4 and IL-13, whereas chronic lesions are dominated by mRNA for IFNγ, IL-5 and IL-12 [16, 17].

Depositions of eosinophil-derived proteins such as major basic protein and eosinophil cationic protein are abundantly present in the dermis. The role of eosinophils in AD is not known. They may be involved in tissue injury.

The cellular interactions involved in the induction of allergic inflammation in AD skin are discussed in the chapters by E. Van Hoffen/F.C. Van Reijsen and E.F. Knol (this volume).

Pathogenesis

The pathogenesis of AD is still a matter of debate. It has been suggested that many exogenous and endogenous factors are responsible for the induction of clinical symptoms. However, the complexity of the disease is dominated by two major features: the genetic influence and association with atopic diseases, pointing to the potential role of allergens, and specific IgE. There is increasing evidence that similar mechanisms are involved in the pathogenesis of atopic diseases, e.g. allergic bronchial asthma, allergic rhinitis and conjunctivitis, and AD.

Allergens

The route by which aeroallergens may induce eczematous skin lesions is still a matter of debate. Tupker [18] and Brinkman [19] showed that eczematous lesions are induced after inhalation of allergens. Some studies show a significant improvement of the eczema after control of indoor allergens [20]. However, so far, there are no parameters (clinical and laboratory) to identify those patients who will succesfully respond. The measures recommended for mite-allergic patients with AD are in general the same as those recommended for asthma or allergic rhinitis (allergen avoidance strategies for mattresses, pillows and carpets).

Skin contact with aeroallergens may also induce eczematous skin lesions 24–48 h after application in patients having allergen specific IgE. This test is named the atopy patch test and proves that allergens may penetrate the skin and induze eczematous responses [21].

The atopy patch test is performed on clinically non-involved skin of the back. The aeroallergen may be administered mixed in petrolatum using the Finn chamber technique or as aqueous solution in a concentration 100 times the concentration used for intradermal testing using a Leucotest skin patch test. The test is read after 24–48 h. The number of positive reactions can be increased by superficially stripping the skin by means of a adhesive tape (10 times stripping).The test shows a high specificity for AD and is negative in normals without atopy. Thus far, the clinical relevance of the test is not known. There are no studies dealing with the correlation of the atopy patch test with the outcome of allergen avoidance studies.

Concerning the relationship between food allergens and AD, many studies have been performed in young children. Infants and young children with AD are frequently sensitized to food allergens such as cow's milk, egg white, peanut, wheat and fish. Almost 60% of children with severe AD have positive food challenges (urticaria, eczema) to one of these foods. In adult patients with AD sensitization to food of animal origin is rare. Occasionally a sensitization to cow's milk, egg or fish may be found. In these cases the clinical reaction to cow's milk may be severe, inducing worsening of the eczema but also of urticaria and oral allergy symptoms. Sensitization to food allergens of plant origin is frequently observed, especially in those patients who are sensitized to birch, grass and mugwort pollen. Evidence has been presented that these pollen allergens share epitopes with food allergens. This cross-reactivity is often widespread and may account for a multitude of positive skin tests to vegetable foods with only a few being of clinical relevance. In explaining the acute often generalized, food-allergic reactions (urticaria, anaphylaxis) that may occur in patients sensitized to birch and mugwort pollen, a careful history, knowledge of the characteristic sensitization patterns and insight into food preparation habits and the components of prepacked foods are usually sufficient to pinpoint the relevant positive skin tests. However, if these clear immediate symptoms are absent in patients with AD, it is often very difficult to judge which food allergens (and to what extent) influence the severity of their skin disease, thus creating a greater need for oral, double-blind, placebo-controlled challenge tests. Apart from being time consuming these oral food challenge tests have other flaws. Food allergy has a variable expression and symptomatology, which may depend on the amount of food eaten, the way it was prepared, and the presence or absence of a combination with alcohol and physical exercise. Furthermore, it is not clear if only single or repeated challenges are necessary to provoke a chronic allergic disease like AD. The question if and how food allergens reach the skin and the mechanism by which they induce allergic inflammation in the skin still has to be answered.

Non-allergic factors

It is well known that sweating may cause itching and secondary eczema in patients with AD. The mechanism, however, is not known. Some patients worsen during the winter months, others during spring or summer. Exposure to sun is generally beneficial, but some patients will experience exacerbation. Seaside or mountain stays may induce a dramatic relief.

Substantial clinical research has been carried out to investigate the skin barrier in AD patients. From these studies, it has become clear that the skin barrier for irritants, such as sodium lauryl sulphate, is decreased. The reason for this stratum corneum defect is not known.

AD is one of the most frequently cited skin disorders with a suspected psychosomatic factor. Daily emotional stress can trigger the itching and scratching. The unpredictable course imposes a great psychological burden on the patient. Evidence for the role of stress in AD comes largely from questionnaire studies and clinical observations. Despite the possible practical relevance of negative social interactions in AD, controlled behavioural observations are lacking.

Complications and diseases associated with atopic dermatitis

Patients with AD have an increased susceptibility for bacterial, viral and mycotic infections. Thus far, it is only speculative whether this is due to a defective cell-mediated immunity or to a defective cell-mediated immunity, or to a defective barrier function of the skin.

Bacterial infections

Normal and lesional skin of patients with AD show a higher colonization with *Staphylococcus aureus*. In some patients specific IgE antibodies to *S. aureus* may occur. Staphylococcal protein A may induce immediate type skin reactions (even in normals). Therefore the pathogenetic relevance of these is not known.

Furuncles, abscesses, erysipelas and systemic signs of infection are uncommon.

Mycotic infections

Pityrosporon ovale (or orbiculare) is a saprophytic lipophilic yeast and belongs to the normal microbial flora of the skin. When it becomes pathogenic, it causes skin lesions, such as pityriasis versicolor. Atopic patients may have positive, immediate-type skin test results with specific IgE antibodies present against Pityrosporon extracts. Pityrosporon species may be involved in the head, neck and shoulder dermatitis. The patients show highly pruritic, inflammatory, eczematous lesions, localized to head, neck and shoulders. Some patients may improve after treatment with topical or systemic antimycotic drugs, but relapses occur frequently after weeks or months.

Viral infections

Herpes simplex

The severity of herpes simplex virus infections varies from localized and mild transient mucocutanous lesions to a widespread and fulminant, potentially lifethreatening disease, eczema herpeticum. It is characterized by the appearance of initially discrete clusters of pruritic vesicles and vesiculopustules at the same stage of development, disseminating within days over a large skin surface area. Fever, malaise and lympadenopathy are often present. The diagnosis may be hampered by the clinical resemblance to acute exacerbations of AD and bacterial superinfections. Both patients with severe AD and patients in clinical remission or minor forms of AD may develop eczema herpeticum. Therapy consists of immediate intravenous antiviral treatment.

Mollusca contagiosa

These are common in children with AD. These are shining, yellowish to pink, umbilicated papules. A thick greasy material can be expressed from the central depression. They show spontaneous healing.

Associated ocular disease

AD may be associated with allergo(rhino)conjunctivitis, blepharoconjunctivitis, atopic or vernal keratoconjunctivitis, ocular herpes simplex infections, keratoconus or retinal detachment.

Atopic blepharoconjunctivitis is difficult to treat. Thickening of the palpebral conjunctiva or a giant papillary hypertrophy is often seen. It may persist for months or years. In rare cases the lesions may lead to ectropion and loss of vision has been described.

Allergic rhinitis and allergic asthma

About 33% of children with AD, at the age of 6 years, have concomittant asthma. Children with severe AD have asthma significantly more often than those with moderate AD. If AD occurs before the age of 3 months the risk of developing asthma is increased.

Allergic rhinitis is relatively rare in young children with AD. Asthma is present in about 50% and allergic rhinitis in about 40% of adult patients with AD. In an individual patient the course of atopic cutaneous or respiratory diseases may be alternating, simultaneous or independent. Some of the AD patients who do not show the clinical symptoms of asthma have an increased airway reactivity following histamine or metacholine challenge. This bronchial hyperreactivity may be independent of the atopic state and seems to parallel the course of AD, especially in those patients who are skin test negative. The mechanism behind this phenomenon is not known [22].

Laboratory tests

No consistent or distinctive laboratory abnormalities are associated with AD. Blood eosinophils are often elevated; there is a direct relation with disease severity. Total serum IgE is increased in around 80% of the patients with AD. SerumIgE levels above 10 000 kU/l may be present. Serum IgE antibodies show a wide variety of allergen specificity.

Serum levels of soluble endothelial adhesion molecules are increased and the level runs parallel with disease activity.

Skin tests with histamine and allergens may differ from normals. The flare response is sometimes decreased. Skin tests should be performed when the patient is in clinical remission.

References

1. Hanifin F, Rajka G (1980) Diagnostic features of atopic dermatitis. *Acta Dermatol Venereol (Stockh)* 92: 44–47

2. Schultz-Larsen F (1986) A genetic epidemiologic study in a population based twin sample. *J Am Acad Dermatol* 15: 487–494

3. Diepgen TL, Fartasch M (1993) Recent epidemiologic and genetic studies in atopic dermatitis. *Acta Dermatol Venereol (Stockh)* 176: 13–21

4. Sandford AJ, Shirikawa T, Moffatt F et al (1993) Localization of atopy and beta subunit of high affinity IgE receptor on chromosome 11q. *Lancet* 241: 332–334

5. Savin JA (1993) Atopy and its inheritance: genetics is building a bridge between the immediate and delayed type components of atopy. *Brit Med J* 307: 1019–1020

6. Coleman R, Trembath RC, Harper JJ et al (1993) Chromosome 11q13 and atopy underlying atopic eczema. *Lancet* 341: 1121–1122

7. Foelster-Holst R, Moises HW, Yang L et al (1998) Linkage between atopy and the IgE high affinity receptor gene at 11q13 in atopic dermatitis families. *Hum Genet* 102: 236–239

8. Cox HE, Moffatt MF, Faux JA et al (1998) Association of atopic dermatitis to the beta subunit of the high affinity immunoglobulin IgE receptor. *Br J Dermatol* 138: 182–187

9. Bieber T, de la Salle H, Wollenberg A et al (1992) Human epidermal Langerhans cells express the high affinity receptor for immunoglobulin IgE. *J Exp Med* 175: 1285–1290

10. Wang B, Rieger A, Kilgus O et al (1992) Epidermal Langerhans cells from normal human skin bind monomeric IgE via Fc-epsilon R1. *J Exp Med* 175: 1353–1365

11. Schultz-Larsen F, Diepgen TL, Svensson A (1996) The occurrence of atopic dermatitis in North-Europe: an international questionnaire study. *J Am Acad Dermatol* 34: 760–764

12. Rajka G (1989) *Essential aspects of atopic dermatitis*. Springer Verlag Berlin

13. Bruijnzeel-Koomen CAFM, van Wichen DF, Toonstra J et al (1986) The presence of IgE molecules on epidermal Langerhans cells frm patients with atopic dermatitis. *Arch Dermatol Res* 278: 199–205

14. Maurer D, Ebner C, Reiniger B et al (1996) The high affinity IgE receptor medaites IgE dependent allergen presentation. *J Immunol* 154: 6285–6290

15. Thepen T, Langeveld-Wildschut EG, Bihari IC et al (1996) Biphasic response against aeroallergens in atopic eczema showing a switch from an initial Th2 response to a Th1 response: an immunocytochemical study. *J Allergy Clin Immunol* 97: 823–827

16. Grewe M, Walther S, Guyfko K et al (1995) Analysis of the cytokine pattern expressed *in situ* in inhalant allergen patch test reactions of atopic dermatitis patients. *J Invest Dermatol* 105: 823–827

17. Hamid Q, Boguniewicz M, Leung DYM (1994) Differential in situ cytokine gene expression in acute versus chronic atopic dermatitis. *J Clin Invest* 94: 870–876

18. Tupker RA, de Monchy JGR, Coenraads PJ et al (1996) Induction of atopic dermatitis by inhalation of house dust mite. *J Allergy Clin Immunol* 97: 1069–1070

19. Brinkman L, Aslander MA, Raaymakers JAM et al 91997) Bronchial and cutaneous

responses in atopic dermatitis patients after allergen inhalation challenge. *Clin Exp Allergy* 27: 1043–1051

20 Tan BB, Weald D, Strickland I et al (1996) Double blind controlled trial of effect of house dust mite allergen avoidance on atopic dermatitis. *Lancet* 347: 15–18

21 Langeveld-Wildschut EG et al (1995) Evaluation of variables influencing the outcome of the atopy patch test. *J Allergy Clin Immunol* 96: 66–73

22 Fabrizi G, Corbo GM, Ferrante E et al (1992) The relationship between allergy clinical symptoms and bronchial responsiveness in atopic dermatitis. *Acta Dermatol Venereol (Stockh)* 176: 68–73

Allergic contact dermatitis: cosmetics

Anton C. de Groot

Department of Dermatology, Carolus-Liduina Ziekenhuis, P.O. Box 1101, 5200 BD 's-Herto-genbosch, The Netherlands

Introduction

Cosmetics are frequent causes of allergic contact dermatitis. After metals, ingredients of cosmetic products are the most common contact allergens identified. Even in the general population, some 2–3% may be allergic to substances that are present in cosmetics and toiletries [1]. The recent introduction of mandatory ingredient labelling in the European Union [2] has greatly facilitated diagnostic procedures in patients suspected to suffer from cosmetic allergy, and the number of allergies identified may yet increase. In addition, labelling now enables these allergic patients to avoid products containing substances that will cause allergic contact dermatitis in them.

In this chapter, some aspects of cosmetic allergy will be discussed, including its epidemiology, clinical picture, the causative products, common cosmetic allergens and diagnostic procedures. A full review of side-effects of cosmetics and toiletries is provided in reference [3]. Other useful information on the subject can be found in references [4–6]. A comprehensive review of adverse reactions to fragrances has also been published [7].

Epidemiology of allergic cosmetic dermatitis

Allergic contact dermatitis from cosmetics and toiletries is far from rare. In the general population, an estimated 1% is allergic to fragrances [8] and 2–3% are allergic to substances that may be present in cosmetics and toiletries [1]. Of dermatological patients patch-tested for suspected allergic contact dermatitis, some 10% are allergic to cosmetic products [1]. Six to 14% of such routinely tested individuals react to the fragrance mix, an indicator allergen for perfume allergy [7]. When tested with 10 popular perfumes, 6.9% of female eczema patients proved to be allergic to them [9] and 3.2% to 4.2% were allergic to fragrances from perfumes present in various cosmetic products [10].

Possibly, the frequency of cosmetic reactions is even underestimated. Many cases are mild and the individual noticing a reaction may simply stop using the causative product and not seek medical attention. In addition, in more than half of all cases, the diagnosis of cosmetic allergy is suspected neither by the patient nor by the doctor [11]. Women are at greater risk of acquiring hypersensitivity to cosmetic ingredients than men because of greater product use.

Clinical picture

The clinical picture of allergic cosmetic dermatitis depends on the type of products used (and consequently, the sites of application) and the degree of the patient's sensitivity. Usually, cosmetics and their ingredients are weak allergens, and the dermatitis resulting from cosmetic allergy is mild: erythema, mild edema, desquamation and papules. Weeping vesicular dermatitis rarely occurs, although some products, especially permanent hair dyes, may cause fierce reactions, notably on the face and ears and less on the scalp. Allergic reactions on the scalp tend to be seborrhoeic dermatitis-like with (temporary) hair loss.

Contact allergy to fragrances may resemble nummular eczema, seborrhoeic dermatitis, sycosis barbae or lupus erythematosus [12]. Lesions in the skin folds may be mistaken for atopic dermatitis. Dermatitis due to perfumes or toilet water tends to be "streaky". Allergy to tosylamide/formaldehyde resin in nail polish may affect the fingers [13], but most allergic reactions are located on the eyelids, in and behind the ears, on the neck, and sometimes around the anus or perivulval area. Eczema of the lips and the perioral region (cheilitis) [14] may be caused by toothpastes, notably from the flavours contained therein [15].

The typical patient suffering from allergic cosmetic dermatitis is a woman aged 20–45 with mild dermatitis of the eyelids. The face itself is also frequently involved and often the dermatitis is limited to the face and/or eyelids. Other predilection sites for cosmetic dermatitis are the neck, the axillae, the arms and the hands. However, all parts of the body may be involved (Tab. 1). Most often, the cosmetics have been applied to previously healthy skin (especially the face), nails or hair. However, allergic cosmetic dermatitis may also be caused by products used on previously damaged skin, for example to treat or prevent dry skin of the arms and legs, or irritant, or atopic hand dermatitis.

The causative products

Most allergic reactions are caused by cosmetics that remain on the skin: "stay-on" or "leave-on" products: skin care products (moisturizing and cleansing creams, lotions, milks, tonics), hair cosmetics (notably hair dyes), nail cosmetics (nail lacquer, nail hardener), deodorants and other perfumes, and facial and eye make-up products [3, 7]. "Rinse-off" or "wash-off" products such as soap, shampoo, bath

Table 1 - Allergic cosmetic dermatitis. Localisation and causative products

Localisation	Causative products
Scalp	Hair dye, permanent wave, hair bleach, shampoo, hair lacquer, hair conditioner, hair gel, styling mousse, hair pomade
Eyelids	Eye shadow, skin care products[1], mascara, eyeliner, dye, eye drops, eye brow pencil, any cosmetic product used on the face, the scalp, the hands or the nails
Ears	Hair dye, nail lacquer, nail hardener, sculptured nails (acrylates)
Face and neck	Skin care products[1], blusher, powder, perfume, aftershave, sunscreens, camouflage stick and cream, any cosmetic product used on the scalp, the hands or the nails
Lips	Lipstick, lip gloss, lip creams and pomades, tooth-paste, mouthwash, contour pencil, nail cosmetics
Axillae	Deodorant, antiperspirant, depilatory
Hands	Skin care products[1], nail lacquer, nail hardener, sculptured nails (fingers), toiletries[2]
Trunk, arms, legs	Skin care products[1], bath and shower products, depilatory (legs), toiletries[2], sunscreens
Genitals	Fragrances (intimate spray), lubricants, vaginal douche
Perianal	Moistened toilet tissues, fragrances, lubricants
Feet	Fragranced powder, deodorant, antiperspirant, nail lacquer (toes)

[1] *Skin care products: creams, lotions, tonics, emulsions used either for moisturizing or cleansing purposes*
[2] *Toiletries: soap, bath foam, shower foam, other foaming products intended for cleansing*

foam, and shower foam rarely elicit or induce contact allergic reactions. This may be explained by the dilution of the product (and, consequently, of the [potential] allergen) under normal circumstances of use, and the fact that the product is removed from the skin by washing after a short period of time. One exeption to this general rule is allergy to the surfactant cocamidopropyl betaine, which has caused many reactions to shampoo in consumers and occupational dermatitis in hairdressers, and to shower gels [4, 16, 17].

The allergens in cosmetic products

Fragrances and preservatives are by far the most common causative ingredients in allergic cosmetic dermatitis [3, 5–7, 11, 18]. Other important allergens are the hair

Table 2 - Common cosmetic allergens: Fragrances [3, 7, 24]

Allergen	Comments	Refs.
Fragrances		
Fragrance mix	Detects < 60–70% of fragrance allergic patients. Frequency of sensitization: 6–14%	[7, 25]

Ingredients of the fragrance mix		
• α-Amylcinnamic aldehyde	Infrequent allergen in the mix	
• Cinnamic alcohol		
• Cinnamic aldehyde	INCI name: cinnamal. Frequent allergen in the mix. Frequency of sensitization: 2,4%	[26] [27]
• Eugenol		
• Geraniol	Infrequent allergen in the mix	
• Hydroxycitronellal		
• Isoeugenol	Frequent allergen in the mix	[28]
• Oak moss absolute	Frequent allergen in the mix	

Other common fragrance allergens	Sensitization > 1% in various studies	[7, 24]
• Benzyl salicylate		
• Cananga oil	INCI name: cananga odorata	
• Citral		
• Coumarin		[29]
• Dehydro-isoeugenol	Present in ylang-ylang oil	
• Dihydrocoumarin		
• Geranium oil		
• Hydroabietyl alcohol		
• Isobornyl cyclohexanol	Synonym: synthetic sandalwood	
• Jasmine absolute/synthetic	INCI name: jasminum officinale	[30]
• Lilial		
• Majantol		
• Methoxycitronellal		
• Methyl heptine carbonate		
• Methyl salicylate		
• Musk ambrette	Photosensitizer	[7]
• Narcissus oil	INCI name: narcissus pseudonarcissus	[25, 30]
• Oil of bergamot	INCI name: citrus bergamia	
• Patchouli oil	INCI name: pogostemon cablin	[25]
• Rose oil	INCI name: rosa	

Table 2 (continued)

Allergen	Comments	Refs.
• Sandalwood oil	INCI name: santalum album	[25, 30]
• Sandela		[25]
• Santalol		
• Ylang-ylang oil	INCI name: cananga odorata	

INCI: International nomenclature cosmetic ingredient [2]

colour *p*-phenylenediamine (and related permanent dyes), the nail lacquer resin tosylamide/formaldehyde resin [13, 19], UV-filters (more often photocontact allergy), and, to a lesser degree, lanolin and its derivatives. Recently emerged cosmetic allergens include the surfactant cocamidopropyl betaine [4, 16, 17] and the preservative methyldibromo glutaronitrile [20]. The vitamin E derivative tocopheryl linoleate caused an epidemic of papular and vesicular (allergic and irritant) contact dermatitis in Switzerland in 1992 [21]. Common cosmetic allergens with references and additional relevant information are listed in Table 2 (fragrances), Table 3 (preservatives) and Table 4 (miscellaneous cosmetic allergens).

Fragrances

Fragrances not only cause allergic contact dermatitis but also irritant reactions, photosensitivity, immediate contact reactions (contact urticaria), pigmented contact dermatitis and (worsening of) respiratory problems. For a full review of side-effects of fragrances (and essential oils) see reference [7]. A recent book on beneficial and adverse reactions to fragrances also provides valuable information [22].

Considering the extensive use of fragrances, the frequency of contact allergy to them is relatively small. In absolute numbers, however, fragrance allergy is common (see the section on epidemiology). Patients allergic to fragrances are usually adult individuals, both women and men. They mainly become allergic by the use of cosmetics and personal care products; occupational contact with fragrances is rarely important [7].

Contact allergy to fragrances usually causes dermatitis of the hands, the face and/or the armpits. Patients appear to become sensitized to fragrances especially by the use of deodorant sprays and/or perfumes, and to a lesser degree by cleansing agents, deodorant sticks or hand lotions [23]. Thereafter, new rashes may appear or are worsened by contact with other fragranced products: other cosmetics, household products, industrial contacts, paper and paper products, laundered fabrics and clothes, topical drugs, and fragrances used as spices in foods and drinks.

Over 100 fragrances have been identified as allergens [7]. Most reactions are caused by the eight fragrances in the perfume mix, and of these oak moss, isoeugenol and cinnamic aldehyde (cinnamal) are the main sensitizers (Tab. 2).

Preservatives

Preservatives [31–33] are added to water-containing cosmetics to inhibit the growth of nonpathogenic and pathogenic micro-organisms, which may cause degradation of the product or endanger the health of the consumer. After fragrances, they are the most frequent cause of allergic cosmetic dermatitis. A list of the most important preservative allergens is provided in Table 3.

Formaldehyde

Formaldehyde is a frequent sensitizer [27] and ubiquitous allergen, with numerous noncosmetic sources of contact. The cosmetic industry uses small but effective concentrations, with the amount of free formaldehyde not exceeding 0.2%, and its use is restricted almost exclusively to rinse-off products. Until recently, most shampoos contained formaldehyde. This practice rarely gave rise to cases of allergic cosmetic dermatitis. In recent years, it has largely been replaced with other preservatives because formaldehyde (when inhaled as gas) is suspected of being a possible human carcinogen.

Formaldehyde donors

Formaldehyde donors are preservatives that, in the presence of water, release formaldehyde. Therefore, cosmetics preserved with such chemicals will contain free formaldehyde, the amount depending on the preservative used, its concentration and the amount of water present in the product. Formaldehyde donors used in cosmetics and toiletries include quaternium-15, imidazolidinyl urea, diazolidinyl urea, 2-bromo-2-nitropropane-1,3-diol, and DMDM hydantoin. In anionic shampoos the amount of formaldehyde released by such donors increases in the order: imidazolidinyl urea < DMDM hydantoin < diazolidinyl urea < quaternium-15. Whereas the use of formaldehyde as a preservative has drastically decreased in recent years, the increased popularity of the formaldehyde donors in the cosmetic industry suggests that an increase in the prevalence of sensitivity to them can be expected [35]. Contact allergy to formaldehyde donors may be due either to the preservative itself or to formaldehyde sensitivity [31, 32].

Table 3 - Common cosmetic allergens: Preservatives

Allergen	Comments	Refs.
Formaldehyde and -donors		
Formaldehyde	Used only in rinse-off products. Frequency of sensitization: 3% [1], 9% [27].	[31, 32]
Formaldehyde donors		
• 2-Bromo-2-nitropropane-1,3-diol	Trade name: Bronopol. Frequency of sensitization: 2.3% [27]. Not a frequent allergen in Europe [35]. Caused many reactions in Eucerin cream in the USA [36].	[34]
• Diazolidinyl urea	Frequency of sensitization: 3.7% [27]. Most allergic patients react to formaldehyde [37]. Cross-reacts to and from imidazolidinyl urea [38]	[37, 38]
• DMDM hydantoin	Frequency of sensitization: 2.3% [27]. No cases of contact allergy to the preservative itself yet reported, although reactions in formaldehyde sensitive subjects are possible [39].	[39]
• Imidazolidinyl urea	Trade name: Germall 115. Frequency of sensitization: 3.1% [27]. Releases little formaldehyde. Cross-reacts to and from diazolidinyl urea [40].	[40]
• Quaternium-15	Trade name: Dowicil 200. Frequency of sensitization: 9.2% [27]. Less frequent allergen in Europe [35, 40]. Half of the reactions are caused by formaldehyde sensitivity [41].	[40, 41]
Other preservatives		
Methyl(chloro)isothiazolinone	Trade names: Kathon CG, Euxyl K 100. Frequency of sensitization: 3% [27]. Decreases [35].	[42, 43]
Methyldibromo glutaronitrile	Trade name: Euxyl K 400. Allergen in cosmetics and (in the Netherlands) in moistened toilet tissues. Frequency of sensitization: see text.	[text]
Parabens	Uncommon allergen in cosmetics. Frequency of sensitization: 1.8% [27]. Reactions are usually caused by topical drugs used on leg ulcers or on eczematous skin.	[33]
Miscellaneous preservatives	Less common allergens: chlorhexidine, chloroacetamide, chloroxylenol, mercurials, quaternary ammonium compounds, sorbic acid, triclosan	[3, 6]

Methyl(chloro)isothiazolinone (MI/MCI, Kathon CG , Euxyl K 100)

MI/MCI is a preservative system containing, as active ingredients, a mixture of methylchloroisothiazolinone and methylisothiazolinone. In recent years, this highly effective preservative has become a major cause of cosmetic allergy in most European countries [42, 43]. Allergic reactions on the face to cosmetics preserved with MCI/MI can have unusual clinical presentations that are very similar to seborrheic dermatitis, lupus erythematosus, lymphocytic infiltrate, photodermatitis or atopic dermatitis [44]. Currently, MCI/MI is mainly used in rinse-off products at low concentrations, which infrequently leads to induction or elicitation of contact allergy [45]. As a consequence, prevalence rates in Europe are decreasing [35].

Methyldibromo glutaronitrile

Euxyl K 400 is a preservative system for cosmetics and toiletries, containing 2 active ingredients: methyldibromo glutaronitrile (synonym: 1,2-dibromo-2,4-dicyanobutane) and phenoxyethanol in a 1:4 ratio. It was thought to be a suitable alternative to the sensitizing MCI/MI, but unfortunately Euxyl K 400 soon proved to be a frequent cause of contact allergy to cosmetics [20] and, in the Netherlands, to moistened toilet tissues [46]. Prevalence rates of sensitization in patients routinely investigated for suspected allergic contact dermatitis were 4% in the Netherlands [46], 2.9% in Italy [47], 2.3% in Germany [48] and 2% [27] to 11.7% in the USA [49].

Other cosmetic allergens

Other allergens in cosmetics that sensitize frequently or occasionally are listed in Table 4.

Diagnostic procedures

The diagnosis of cosmetic allergy [3, 5] should strongly be suspected in any patient presenting with dermatitis of the face, eyelids, lips and neck. Cosmetic allergic dermatitis may develop on previously healthy skin of the face or on already damaged skin (irritant contact dermatitis, atopic dermatitis, seborrheic dermatitis, allergic contact dermatitis from other sources). Also, dermatitis of the arms and hands may be caused or worsened by skin care products to treat or prevent dry skin, irritant or atopic dermatitis. Patchy dermatitis in the neck and around the eyes is suggestive of cosmetic allergy from nail lacquers or hardeners. More widespread problems may be caused by ingredients in products intended for general application to the body. Hypersensitivity to other products such as deodorants usually causes a reaction localised to the site of application. A thorough history of cosmetic usage should

Table 4 - Common cosmetic allergens: Miscellaneous

Allergen	Comments	Refs.
Antioxidants	Occasional causes of cosmetic allergy: BHA, BHT, *t*-butylhydroquinone, nordihydroguiaretic acid, gallates (dodecyl, octyl, propyl), and tocopherol (vitamin E).	[50, 51] [52, 53] [27, 54]

Emulsifiers, humectants, emollients and surfactants

Allergen	Comments	Refs.
Cocamidopropyl betaine	Amphoteric surfactant in shampoos and bath/ shower products. Prevalence rates of sensitiza- tion: 3.7%–5% [55–57]. Allergen may be di- methylaminopropylamine [55] and/or cocamido- propyl betaine [58]. Occupational hazard to hair- dressers. Risk of false-positive patch test reactions.	[16, 17]
Lanolin and derivatives	Very widely used as emollients and emulsifiers. INCI name: lanolin alcohol. Frequency of sen- sitization: 3.3% [27]. Most reactions are caused by their presence in topical pharmaceutical pre- parations [59]. Risk of sensitization in cosmetics is small [61, 62].	[61, 62]
Oleamidopropyl dimethylamine	Formerly an important allergen in the Nether- lands. Reactions are still found, but may be caused by the impurity dimethylaminopropyl- amine (as in cocamidopropyl betaine).	[63]
Propylene glycol	Very widely used, also in non-cosmetic formu- lations. Frequency of sensitization: 1.1% [27]. Formerly incorrectly thought to be an important allergen.	[64, 65]

Hair dyes

Allergen	Comments	Refs.
p-Phenylenediamine (PPD)	PPD and related dyes are important sensitizers. Frequency of sensitization: 6.8% [27]. Occupa- tional hazard to hairdressers and beauticians [67]. Other allergenic dyes: *m*-aminophenol, *p*-amino- phenol, 2-nitro-*p*-phenylenediamine, N-phenyl- *p*-phenylenediamine, toluene-2,4/2,5-diamine.	[54, 66]

Allergen	Comments	Refs.
Nail lacquer allergens	The main allergen is tosylamide/formaldehyde resin. Frequency of sensitization: 1.6% [27], in women habitually using it 6.6% [70]. Localisation	[68, 69]

Table 4 (continued)

Allergen	Comments	Refs.
	of dermatitis: 80% neck and face (eyelids), periungual [13], sometimes genital area and trunk. Other less frequent allergens: see [71, 72].	
Ultraviolet light (UV) filters	Sunscreens, also incorporated in face cosmetics. Sunscreen preparations cause irritation in 15% of users [75]. Photocontact allergy more frequent than contact allergy. Is seen especially in habitual users with photodermatoses [76]. Most reactions are caused by PABA (USA), dibenzoylmethanes (isopropyl dibenzoylmethane, withdrawn in 1993) and benzophenones: benzophenone-3 (oxybenzone).	[54, 73] [74]
Miscellaneous allergens	Cause cosmetic allergy occasionally	
• Colophonium (rosin)	Film former. In lipsticks and eye makeup	[77]
• Dihydroxyacetone	Self-tanning agent	[54]
• Glyceryl thioglycolate	Permanent wave. Occupational hazard for hairdressers. Frequency of sensitization: 5% [27]	[54, 67]
• Kojic acid	Depigmenting agent. Allergen in Japan	[78]
• Phenyl salicylate	In lipsticks	
• Propolis	Natural ingredient	[79]
• Tocopheryl linoleate	Epidemic of irritant and allergic reactions in Switzerland in 1992	[21]

always be obtained. When the diagnosis of cosmetic allergy is suspected, patch tests should be performed to confirm the diagnosis and identify the sensitizer. Only in this manner can the patient be counselled about future use of cosmetic (and other) products and the prevention of recurrences of dermatitis from cosmetic or non-cosmetic sources. Patch tests should be performed with the NACDG or European routine series, a "cosmetic series" containing known cosmetic allergens , and all products used by the patient. The European routine series contains a number of cosmetic allergens and "indicator" allergens: colophonium, balsam of Peru (INCI name: Myroxylon Pereirae), the fragrance-mix, formaldehyde, quaternium-15, methyl (chloro)-isothiazolinone, wool alcohols (INCI name: lanolin alcohol) and *p*-phenyl-enediamine. A suggested "cosmetic series" is shown in Table 5. The patient's products should always be tested, but false-positive and false-negative reactions occur frequently.

In certain cases, allergy to cosmetics is strongly suspected, but patch testing remains negative. In such patients, repeated open application tests (ROAT) and/or usage tests can be performed. In the ROAT, the product is applied twice daily for a maximum of 14 days to the antecubital fossa. A negative reaction after 2 weeks makes sensitivity highly unlikely. This procedure should be performed with all suspected products except detergent-containing cosmetics such as soap, shampoo and shower foam.

In the usage test, all cosmetic products are stopped, until the dermatitis has disappeared. Then, cosmetics are reintroduced as normally used, one at a time, with an interval of 3 days for each product, until a reaction develops.

Diagnosing fragrance allergy

A perfume may contain as many as 200 or more individual ingredients. This makes the diagnosis of perfume allergy by patch test procedures complicated. The fragrance mix, or perfume mix, was introduced as a screening tool for fragrance sensitivity in the late 1970s [80]. It contains eight commonly used fragrances: α-amyl-cinnamic aldehyde, cinnamic alcohol, cinnamal (cinnamic aldehyde), eugenol, geraniol, hydroxycitronellal, isoeugenol and oak moss absolute. Between 6% and 14% [7] of patients routinely tested for suspected allergic contact dermatitis react to it. It is estimated that this mix detects 70–80% of all cases of fragrance sensitivity; this may be an overestimation, as it was positive in only 57% of patients who were allergic to popular commercial fragrances [9] and in < 50% in fragrance allergic patients in a world-wide study [25].

The finding of a positive reaction to the fragrance mix should be followed by a search for its relevance, i.e. is fragrance allergy the cause of the patient's current or previous complaints or does is at least contribute to it. Often, however, correlation with the clinical picture is lacking and many patients can tolerate perfumes and fragranced products without problem [7]. This may sometimes be explained by irritant (false positive) patch test reactions to the mix. Alternative explanations include the absence of relevant allergens in those products or a concentration too low to elicit clinically visible allergic contact reactions.

It is assumed that between 50% and 65% of all positive patch test reactions to the mix are relevant, although this is sometimes hard to prove. Nevertheless, there is a highly significant association between the occurrence of self-reported visible skin symptoms to scented products earlier in life and a positive patch test to the fragrance mix, and most fragrance sensitive patients are aware that the use of scented products may cause skin problems [81].

Table 5 - Suggested allergens for a "cosmetic screening series" [6]

Allergen	Function	Test conc. and vehicle
Amerchol L 101[3]	emulsifier	50% pet
Benzophenone-3 (oxybenzone)	sunscreen	2% pet[5]
Benzophenone-10 (mexenone)	sunscreen	2% pet[5]
BHA (butylated hydroxyanisole)	antioxidant	2% pet
BHT (butylated hydroxytoluene)	antioxidant	2% pet
2-Bromo-2-nitropropane-1,3-diol	preservative	0.5% pet
Cetearyl alcohol[4]	emulsifier	30% pet
Cocamidopropyl betaine	surfactant	1% aqua
Diazolidinyl urea[2]	preservative	2% aqua or pet
Fragrance-mix[1]	fragrance	8 × 1% pet
Glyceryl thioglycolate	permanent waving agent	1% pet
Imidazolidinyl urea[2]	preservative	2% pet
Methyl(chloro)isothiazolinone[1]	preservative	100 ppm in water
Methyldibromo glutaronitrile	preservative	0.5% pet
Octyl dimethyl PABA	sunscreen	2% pet[5]
PABA	sunscreen	2% pet[5]
Parabens[1]	preservatives	5 × 3% pet
Propolis	natural ingredient	10% pet
Propylene glycol	humectant	10% water
Tosylamide/formaldehyde resin	nail lacquer resin	10% pet

[1]*present in the European standard series*
[2]*present in the NACDG (North American Contact Dermatitis Group) series*
[3]*INCI name: lanolin alcohol and paraffinum liquidum*
[4]*INCI name: cetyl alcohol, stearyl alcohol*
[5]*recent data suggest that the test concentrations of UV filters should be increased to up to 10%*

In perfume mix allergic patients with concomitant positive reactions to perfumes or scented products used by them, interpretation of the reaction as relevant is highly likely. In such patients the incriminated cosmetics very often contain fragrances present in the mix and thus, the fragrance mix appears to be a good reflection of actual exposure [82]. Indeed, one or more of the ingredients of the mix are present in nearly all deodorants [83], popular prestige perfumes [9], perfumes used in the formulation of other cosmetic products [10] and natural ingredient based cosmetics [84], often in levels high enough to cause allergic reactions [26, 28]. Thus, fragrance allergens are ubiquitous and virtually impossible to avoid if perfumed cosmetics are used.

References

1   De Groot AC (1990) Labelling cosmetics with their ingredients. *Br Med J* 300: 1636–1638

2   De Groot AC, Weyland JW (1997) Conversion of common names of cosmetic allergens to the INCI nomenclature. *Contact Dermatitis* 37: 145–150

3   De Groot AC, Weyland JW, Nater JP (1994) *Unwanted effects of cosmetics and drugs used in dermatology*, 3rd ed. Elsevier, Amsterdam

4   De Groot AC (1997) Contact allergens. What's new? Cosmetic dermatitis. *Clin Dermatol* 15: 485–492

5   De Groot AC (1998) Fatal attractiveness: The shady side of cosmetics. *Clin Dermatol* 16: 167–179

6   De Groot AC, White IR (2000) Cosmetics and skin care products. In: R Rycroft, T Menné, P Frosch, J-P Lepoittevin (eds): *Textbook of contact dermatitis*, 3rd ed. Springer-Verlag, Heidelberg; *in press*

7   De Groot AC, Frosch PJ (1997) Adverse reactions to fragrances. A clinical review. *Contact Dermatitis* 36: 57–86

8   Nielsen NH, Menné T (1992) Allergic contact sensitization in an unselected Danish population. *Acta Derm Venereol* 72: 456–460

9   Johansen JD, Rastogi SC, Menné T (1996) Contact allergy to popular perfumes; assessed by patch test, use test and chemical analysis. *Br J Dermatol* 135: 419–422

10 Johansen JD, Rastogi SC, Andersen KE, Menné T (1997) Content and reactivity to product perfumes in fragrance mix positive and negative eczema patients. A study of perfumes used in toiletries and skin-care products. *Contact Dermatitis* 36: 291–296

11 Adams RM, Maibach HI (1985) A five-year study of cosmetic reactions. *J Am Acad Dermatol* 13: 1062–1069

12 Meynadier J-M, Raison-Peyron N, Meunier L, Meynadier J (1997) Allergie aux parfums. *Rev fr Allergol* 37: 641–650

13 Lidén C, Berg M, Färm G (1993) Nail varnish allergy with far-reaching consequences. *Brit J Dermatol* 128: 57–62

14 Ophaswongse S, Maibach HI (1995) Allergic contact cheilitis. *Contact Dermatitis* 33: 365–370

15 Sainio E-L, Kanerva L (1995) Contact allergens in toothpastes and a review of their hypersensitivity. *Contact Dermatitis* 33: 100–105

16 De Groot AC (1997) Cocamidopropyl betaine: A "new" important cosmetic allergen. *Dermatosen* 45: 60–63

17 De Groot AC, van der Walle HB, Weijland JW (1995) Contact allergy to cocamidopropyl betaine. *Contact Dermatitis* 33: 419–422

18 De Groot AC, Bruynzeel DP, Bos JD, van der Meeren HLM, van Joost T, Jagtman BA, Weyland JW (1988) The allergens in cosmetics. *Arch Dermatol* 124: 1525–1529

19 Berne B, Boström Å, Grahnén AF, Tammela M (1996) Adverse effects of cosmetics and

toiletries reported to the Swedish Medical Product Agency 1989–1994. *Contact Dermatitis* 34: 359–362

20 De Groot AC, van Ginkel CJW, Weyland JW (1996) Methyldibromo glutaronitrile (Euxyl K 400): An important "new" allergen in cosmetics. *J Am Acad Dermatology* 35: 743–7

21 Wyss M, Elsner P, Homberger H-P, Greco P, Gloor M, Burg G (1997) Follikuläres Kontaktekzem auf eine Tocopherol-linoleat-haltige Körpermilch. *Dermatosen* 45: 25–28

22 Frosch PJ, Johansen JD, White IR (eds) (1998) *Fragrances. Beneficial and adverse effects.* Springer-Verlag, Berlin

23 Johansen JD, Andersen TF, Kjøller M, Veien N, Avnstorp C, Andersen KE, Menné T (1998) Identification of risk products for fragrance contact allergy: A case-referent study based on patients' histories. *Am J Contact Dermatitis* 9: 80–87

24 De Groot AC (2000) Dermatological problems linked to perfumes. In: AO Barel, HI Maibach, M Paye (eds): *Handbook of cosmetic science and technology.* Marcel Dekker, New York; *in press*

25 Larsen W, Nakayama H, Lindberg M, Fischer T, Elsner P, Burrows D, Jordan W, Shaw S, Wilkinson J, Marks J Jr et al (1996) Fragrance contact dermatitis: a worldwide multicenter investigation (part 1). *Am J Contact Dermatitis* 7: 77–83

26 Johansen JD, Andersen KE, Rastogi SC, Menné T (1996) Threshold responses in cinnamic-aldehyde-sensitive subjects: results and methodological aspects. *Contact Dermatitis* 34: 165–171

27 Marks JG, Jr, Belsito DV, DeLeo VA, Fowler JF Jr, Fransway AF, Maibach HI (1998) North American Contact Dermatitis Group patch test results for the detection of delayed-type hypersensitivity to topical allergens. *J Am Acad Dermatol* 38: 911–918

28 Johansen JD, Andersen KE, Menné T (1996) Quantitative aspects of isoeugenol contact allergy assessed by use and patch tests. *Contact Dermatitis* 34: 414–418

29 Kunkeler ACM, Weijland JW, Bruynzeel DP (1998) The role of coumarin in patch testing. *Contact Dermatitis* 39: 327–328

30 Larsen W, Nakayama H, Fischer T, Elsner P, Frosch P, Burrows D, Jordan W, Shaw S, Wilkinson J, Marks J Jr et al (1998) A study of new fragrance mixtures. *Am J Contact Dermatitis* 9: 202–206

31 Fransway AF (1991) The problem of preservation in the 1990s. Statement of the problem. Solution(s) of the industry and the current usage of formaldehyde and formaldehyde-releasing biocides. *Am J Contact Dermatitis* 2: 6–23

32 Fransway AF, Schmitz NA (1991) The problem of preservation in the 1990s. II. Formaldehyde and formaldehyde-releasing biocides: Incidences of cross-reactivity and the significance of the positive response to formaldehyde. *Am J Contact Dermatitis* 2: 78–88

33 Fransway AF (1991) The problem of preservation in the 1990s. III. Agents with preservative function independent of formaldehyde release. *Am J Contact Dermatitis* 2: 145–174

34   Frosch PJ, White IR, Rycroft RJG, Lahti A, Burrows D, Camarasa JG, Ducombs G, Wilkinson JD (1990) Contact allergy to Bronopol. *Contact Dermatitis* 22: 24–26

35   Goossens A, Claes L, Drieghe J, Put E (1997) Antimicrobials: preservatives, antiseptics and disinfectants. *Contact Dermatitis* 39: 133–134

36   Storrs F, Bell DE (1983) Allergic contact dermatitis to 2-bromo-2-nitropane-1,3-diol in a hydrophilic ointment. *J Am Acad Dermatol* 8: 157–164

37   Hectorne KJ, Fransway AF (1994) Diazolidinyl urea: incidence of sensitivity, patterns of cross-reactivity and clinical relevance. *Contact Dermatitis* 30: 16–19

38   De Groot AC, Bruynzeel DP, Jagtman BA, Weyland JW (1988) Contact allergy to diazolidinyl urea (Germall II). *Contact Dermatitis* 18: 202–205

39   De Groot AC, van Joost T, Bos JD, van der Meeren HLM, Weyland JW (1988) Patch test reactivity to DMDM hydantoin. Relationship to formaldehyde. *Contact Dermatitis* 18: 197–201

40   Jacobs M-C, White IR, Rycroft RJG (1995) Patch testing with preservatives at St. John's from 1982 to 1993. *Contact Dermatitis* 33: 247–254

41   Parker LU, Taylor JS (1991) A 5-year study of contact allergy to quaternium-15. *Am J Contact Dermatitis* 2: 231–234

42   De Groot AC, Weijland JW (1988) Kathon CG: A review. *J Am Acad Dermatol* 18: 350–358

43   De Groot AC (1990) Methylisothiazolinone/methylchloroisothiazolinone (Kathon CG) allergy: an updated review. *Am J Contact Dermatitis* 1: 151–6

44   Morren M-A, Dooms-Goossens A, Delabie J, De Wolf-Peeters Ch, Marien K, Degreef H (1992) Contact allergy to isothiazolinone derivatives: unusual clinical presentations. *Dermatology* 184: 260–264

45   Frosch PJ, Hannuksela M, Andersen KE, Wilkinson JD, Shaw S, Lachapelle JM (1995) Chloromethylisothiazolinone/methylisothiazolinone (CMI/MI) use test with a shampoo on patch-test-positive subjects. *Contact Dermatitis* 32: 210–217

46   De Groot AC, de Cock PAJJM, Coenraads PJ (1996) Methyldibromo glutaronitrile is an important contact allergen in the Netherlands. *Contact Dermatitis* 34: 118–20

47   Tosti A, Vincenzi C, Trevisi P, Guerra L (1995). Euxyl K 400: incidence of sensitization, patch test concentration and vehicle. *Contact Dermatitis* 33: 193–5

48   Schnuch A, Geier J (1994) Die häufigsten Kontaktallergene im zweiten Halbjahr 1993. *Dermatosen* 42: 210–211

49   Jackson JM, Fowler JF (1998) Methyldibromoglutaronitrile (Euxyl K400): A new and important sensitizer in the United States? *J Am Acad Dermatol* 38: 934–937

50   White IR, Lovell CR, Cronin E (1984) Antioxidants in cosmetics. *Contact Dermatitis* 11: 265–267

51   Le Coz CJ, Schneider G-A (1998) *Contact Dermatitis* from tertiary-butylhydroquinone in a hair dye, with cross-sensitivity to BHA and BHT. *Contact Dermatitis* 39: 39–40

52   Serra-Baldrich E, Puig LL, Gimenez Arnau A, Camarasa JG (1995) Lipstick allergic contact dermatitis from gallates. *Contact Dermatitis* 32: 359–360

53   Parsad D, Saini R, Verma N (1997) Xanthomatous reaction following contact dermatitis from vitamin E. *Contact Dermatitis* 37: 294

54   Goossens A, Beck MH, Haneke E, McFadden JP, Nolting S, Durupt G, Ries G (1999) Adverse cutaneous reactions to cosmetics. *Contact Dermatitis* 40: 112–113

55   Pigatto PD, Bigardi AS, Cusano F (1995) Contact dermatitis to cocamidopropylbetaine is caused by residual amines: relevance, clinical characteristics, and review of the literature. *Am J Contact Dermatitis* 6: 13–16

56   Fowler JF (1993) Cocamidopropyl betaine: the significance of positive patch test results in twelve patients. *Cutis* 52: 281–284

57   Angelini G, Foti C, Rigano L (1995) 3-Dimethylaminopropylamine: a key substance in contact allergy to cocamidopropylbetaine? *Contact Dermatitis* 32: 96–9

58   Fowler JF, Fowler LM, Hunter JE (1997) Allergy to cocamidopropyl betaine may be due to amidoamine: a patch test and product use test study. *Contact Dermatitis* 37: 276–281

59   Wilson CI, Cameron J, Powell SM, Cherry G, Ryan TJ (1997) High incidence of contact dermatitis in leg-ulcer patients – implications for management. *Clin Exp Dermatol* 16: 250–261

60   Nachbar F, Korting HC, Plewig G (1993) Zu Bedeutung des positiven Epikutantests auf Lanolin. *Dermatosen* 41: 227–236

61   Kligman AM (1998) The myth of lanolin allergy. *Contact Dermatitis* 39: 103–107

62   Wolf R (1996) The lanolin paradox. *Dermatology* 192: 198–202

63   Foti C, Rigano L, Vena GA, Grandolfo M, Liguori G, Angelini G (1995) Contact allergy to oleamidopropyl dimethylamine and related substances. *Contact Dermatitis* 33: 132–133

64   Funk JO, Maibach HI (1994) Propylene glycol dermatitis: re-evaluation of an old problem. *Contact Dermatitis* 31: 236–241

65   Wahlberg JE (1994) Propylene glycol: Search for a proper and nonirritant patch test preparation. *Am J Contact Dermatitis* 5: 156–159

66   Marcoux D, Riboulet-Delmas G (1994) Efficacy and safety of hair-coloring agents. *Am J Contact Dermatitis* 5: 123–129

67   Conde-Salazar L, Baz M, Guimaraens D (1995) *Contact Dermatitis* in hairdressers: patch test results in 379 hairdressers. *Am J Contact Dermatitis* 6: 19–23

68   Rosenzweig R, Scher RK (1993) Nail cosmetics: Adverse reactions. *Am J Contact Dermatitis* 4: 71–77

69   Barnett JM, Scher RK (1992) Nail cosmetics. *Int J Dermatol* 31: 675–681

70   Tosti A, Guerra L, Vincenzi C (1993) Contact sensitization caused by toluene sulfonamide-formaldehyde resin in women who use nail cosmetics. *Am J Contact Dermatitis* 4: 150–153

71   Hausen BM (1994) Nagellack-Allergie. *H+G Z Hautkr* 69: 252–262

72   Hausen BM, Milbrodt M, Koenig WA (1995) The allergens of nail polish (I). Allergenic constituents of common nail polish and toluenesulfonamide-formaldehyde resin (TS-F-R). *Contact Dermatitis* 33: 157–164

73   Funk JO, Dromgoole SH, Maibach HI (1995) Sunscreen intolerance. Contact sensitiza-

tion, photocontact sensitization, and irritancy of sunscreen agents. *Clin Dermatol* 13: 473–481

74 Gonçalo M, Ruas E, Figueiredo A, Gonçalo S (1995) Contact and photocontact sensitivity to sunscreens. *Contact Dermatitis* 33: 278–280

75 Foley P, Nixon R, Marks R, Frowen K, Thompson S (1993) The frequency of reactions to sunscreens: results of a longitudinal population-based study on the regular use of sunscreens in Australia. *Br J Dermatol* 128: 512–518

76 Bilsland D, Ferguson J (1993) Contact allergy to sunscreen chemicals in photosensitivity dermatitis/actinic reticuloid syndrome (PD/AR) and polymorphic light eruption. *Contact Dermatitis* 29: 70–73

77 Batta K, Bourke JF, Foulds IS (1997) Allergic contact dermatitis from colophony in lipsticks. *Contact Dermatitis* 36: 171–172

78 Nakagawa M, Kawai K, Kawai K (1995) Contact allergy to kojic acid in skin care products. *Contact Dermatitis* 32: 9–13

79 Hausen BM, Wollenweber E, Senff H, Post B (1987) Propolis allergy (I). Origin, properties, usage and literature review. *Contact Dermatitis* 17: 163–170

80 Larsen WG, Nethercott JR (1997) Fragrances. *Clin Dermatol* 15: 499–504

81 Johansen JD, Andersen TF, Veien N, Avnstorp C, Andersen KE, Menné T (1997) Patch testing with markers of fragrance contact allergy. Do clinical tests correspond to patients' self-reported problems? *Acta Derm Venereol (Stockh)* 77: 149–153

82 Johansen JD, Rastogi SC, Menné T (1996) Exposure to selected fragrance materials. A case study of fragrance-mix-positive eczema patients. *Contact Dermatitis* 34: 106–110

83 Rastogi SC, Johansen JD, Frosch PJ, Menné T, Bruze M, Lepoittevin JP, Dreier B, Andersen KE, White IR (1998) Deodorants on the European market: quantitative chemical analysis of 21 fragrances. *Contact Dermatitis* 38: 29–35

84 Rastogi S, Johansen JD, Menné T (1996) Natural ingredients based cosmetics. Content of selected fragrance sensitizers. *Contact Dermatitis* 34: 423–426

Clinical aspects of occupational contact dermatitis

Cornelis J.W. van Ginkel

Department of Dermatology/Allergology, University Hospital, P.O. Box 85500, 3508 GA Utrecht, Netherlands

Introduction

Skin disease constitutes about one-quarter of all occupational diseases. Pulmonary problems, muscle skeletal disorders and mental illness comprise the great majority of the other occupational diseases. Of the occupational dermatoses contact dermatitis is the most important. In this chapter the following aspects of contact dermatitis are discussed: etiology, clinical features, analysis and finally treatment and prevention. The emphasis is on clinical approach, illustrated through several cases.

Etiology

The pathogenesis of contact dermatitis may be theoretically classified as irritant or allergic. However, in daily practice almost all cases of occupational contact dermatitis comprise some degree of irritancy due to environmental factors. Table 1 summarizes the most common occupational irritant substances. Only a minority (about 25%) of cases is dominated by a delayed type hypersensitivity (type IV allergy) to one or more occupational chemicals. For example, many hairdresser apprentices have hand eczema due to exposure to water and shampoo; only in some is the irritant contact dermatitis superimposed by sensitization to, for example, acid permanent wave solution or hair dye. Sometimes, the hand eczema is essentially atopic dermatits, challenged by occupational irritant substances. Meticulous medical history will reveal atopic dermatitis in flexures during childhood. The prognosis of this occupationally aggravated atopic dermatitis is generally poor. Notably in food handling (bakers, cooks, butchers, etc.) the hand dermatitis can be mediated by an IgE-response to food proteins. The first complaints will be an itching wheal-and-flare reaction upon contact with food (contact urticaria). Later, the clinical picture will evolve into dermatitis (so-called protein contact dermatitis) upon continuing re-exposure to food (see Case 1). In Table 2 the food ingredients are reviewed which can induce an IgE-response, notably in patients with an atopic constitution.

Immunology and Drug Therapy of Allergic Skin Diseases, edited by C.A.F.M. Bruijnzeel-Koomen and E.F. Knol

Table 1 - Irritant occupational substances[1]

1 Water
2 Soap[2]
3 Juice[3]
4 Mild alkalis/acids[4]
5 Organic solvents

[1]*Irritant physical factors are not discussed such as ambient temperature/humidity, radiation and mechanical noxae (pressure, friction, occlusion, sharp particles).*
[2]*Synonymous with detergent, surfactant, emulsifier.*
[3]*From fresh fruit, vegetables, meat and fish. Fruit such as kiwi, pineapple and papaya contain high levels of proteolytic activity.*
[4]*Strong acids/alkalis and chemicals such as hydrofluoric acid, ethylene oxide etc. will cause chemical burns upon accidental exposure. This effect will not be discussed.*

Case 1
A 45-year old male presented with hand dermatitis of many years' duration. The dermatitis, initially only contact urticaria, was time-related to his job of removing the innards from fresh eels prior to smoking them. Skin prick test with the unprocessed eel was strongly positive, whereas this test was negative with smoked eel. Changing his job to filleting smoked eels resolved this longstanding dermatitis.

Clinical features

Localization

Occupational contact dermatitis affects predominantly the hands (fingertips and dorsa) and forearms. With volatile chemicals the face and ears are also involved ("air-borne contact dermatitis"), sometimes exclusively (see Case 2). Due to poor toilet hygiene, the genito-anal region can also be affected since the skin on these sites is rather vulnerable.

Time-relationship

Basically occupational dermatitis is time-related to the job. During long absences from work the dermatitis will improve or even resolve. Often, a weekend break is

Table 2 - Food products which can induce IgE-mediated contact urticaria/protein contact dermatitis

Vegetable	Animal
fruit	shell-fish
potatoes	fish
celery/carrots	chicken
lettuce	egg
onion/garlic	
flour/dough	

Case 2

A 32-year old, male fork-lift operator developed an itching dermatitis on his ears and to a lesser extent on his forehead. Patch-testing revealed positive patches for mercapto-benzothiazole (a black rubber additive) and for an ether-extracted sample of black-colored dust, collected from the floor of the workplace. Apparently, the dust emanated from the tires of the fork-lift truck.

Case 3

A 28-year old female who worked for many years with cutflowers and potted plants (both culturing and selling) without skin problems, suddenly developed a blistering swelling of hands, forearms and face, disabling her for some weeks. Patch-testing showed a strong sensitization to primin. Apparently, her employer had recently added the old-fashioned conical primrose (*Primula obconica*, the main source of primin) to their assortment.

---

not sufficient to get a significant improvement. It is important to realize that construction workers and painters sometimes do odd jobs during their holidays or even sick leave with consequently no improvement of their dermatitis. Changes in working conditions and introduction of new, irritant or allergic substances will obscure a clearcut time relationship (see Case 3).

Traditional, small-scale occupations

Occupational contact dermatitis mainly occurs in rather traditional occupations in a small-scale working environment. Under these circumstances the workers still

Table 3 - High-risk occupations

Occupation	Allergen
Construction worker	chromate, epoxy resin, biocides
Machinist	biocides, corrosion inhibitors
Hairdresser	p-phenylenediamine
	glyceryl thioglycolate
	nickel
	ammonium persulfate
Florist	sesquiterpene lactones
	tulipaline
Food worker	see Table 2

have frequent and intense manual contact with a variety of chemicals, resulting in a high prevalence of contact dermatitis. Such contact is absent in large-scale industries such as oil, steel, bulk chemicals, food products, cars, etc. Table 3 summarizes some of these traditional occupations. For reasons of time other high-risk occupations have been deleted such as medical and dental workers, veterinarians, farmers, horticulturists, electronic workers and housekeeping people.

The most important allergens are included in this table. In the following paragraph these occupations will be briefly discussed:

(1) *Construction worker.* Masons, jointers, carpenters, plumbers, tilesetters, plasterers and painters will encounter the various listed allergens depending on the job. With water-based products like paints and plaster the worker will be in contact with the preservatives, whereas chromate is contained only in fresh cement/concrete (see Case 4).

(2) *Machinist.* Metalworking fluids (cutting oils) are essential for cooling and lubrication in metal-cutting. These fluids are emulsions of mineral oils in water. Because of their content of emulsifiers and preservatives (biocides) they can cause both irritant and allergic contact dermatitis.

(3) *Hairdresser.* Many hairdresser apprentices are affected by hand dermatitis, due to irritancy by frequent shampooing. With experienced hair-dressers the hand dermatitis is often due to allergy. The putative allergens are p-phenylene diamine and its derivatives (hair dyes), glyceryl thioglycolate (acid permanent wave), nickel (scissors, clips, pins and rollers) and ammonium persulfate (bleaching agent).

(4) *Florist.* If patchtesting in a florist with handdermatitis reveals a positive result for sesquiterpene lactones (SL's) mix, the prognosis is very poor. SL's are present

---

Case 4

In a 37-year-old tilesetter a hand dermatitis proved to be caused by allergy to epoxy resins. Restricting his daily task to cutting tiles resolved the hand dermatitis, but the facial dermatitis persisted due to the prevalence of epoxy resins in the work site. It was necessary for him to quit his job.

Case 5

A 57-year-old former florist, recently incapacitated due to an allergy to sesquiterpene lactones, had a flare-up of his hand and facial dermatitis. A detailed history taking revealed his new pastime: helping his wife in the kitchen, including cutting SL-rich vegetables like lettuce, endive and french endive.

---

in almost all composite cutflowers ("daisy family") and these cutflowers comprise about one half to three-quarters of a normal assortment. The prognosis of the SL-positive florist is generally incapacitation. However, if testing shows a positive patch for tulipaline, the allergen from the peruvian lily (alstroemeria) and tulip bulbs, strict avoidance of contact with these cutflowers, less than one tenth of the assortment, will resolve the dermatitis (see Case 5).

(5) *Food worker.* This category comprises a variety of occupations such as baker, butcher, green grocer, fishmonger, cook and kitchen-assistant etc. In addition to irritancy, type-I allergy to food ingredients may dominate the hand dermatitis (see also Table 2).

Analysis/work up

To trace the cause of contact dermatitis in an individual worker a three-step procedure is needed:

Step 1: *Information about exposure.* It is essential to get good insight into the working environment and to collect data about the exposure to chemical and mechanical factors. Safety data sheets from manufacturers are useful but seldom sufficient. To get detailed data regarding the chemical identity, for example, of the biocides in metalworking fluids is a tedious and sometimes frustrating job. Although time-consuming, it is often very useful to visit the work site.

Step 2: *Patch-testing.* Patch-testing has to be done with the European Standard Series, supplemented with batteries for specific occupations like hairdresser, machinist, florist, industrial designer, etc. One has to be very restrictive in testing samples, brought from the workplace. Testing is only allowable if chemical identity and concentration of the ingredients of the sample are known. Testing samples of unknown composition bear the risk of heavy local irritation (even burn ulcers) and/or systemic intoxication.

Step 3: *Adjustment of the working environment.* When the putative allergen has been traced, substitution or deletion of that allergen from the working environment is the next step, followed by evaluation of this action. For example, for a florist with hand dermatitis with a positive patch test for tulipaline, it is likely that the hand dermatitis will disappear when the alstroemerias are deleted from his assortment. Another example: A machinist with a longstanding hand dermatitis proved to be sensitized to MCI/MI, a biocide present in the cutting oil of his lathe. The next step is substitution of the cutting oil with one free of this biocide to see if the hand dermatitis will clear up. Sometimes the result is very disappointing if the hand dermatitis is mainly due to irritant factors or the dermatitis is essentially an aggravation of latent atopic dermatitis. In such cases the sensitization to a specific allergen has to be interpreted as an epiphenomenon with only slight clinical relevance.

Treatment and prevention

Tracing the cause of the contact dermatitis is the core of the treatment. Reduction of exposure to irritants and allergens is the baseline of the treatment. Consistent use of proper gloves is an easy and low-cost first-line approach. It is important that the appropriate type of gloves be chosen. For example, vinyl gloves are useless for a dental worker who is sensitized to acrylates, as vinyl gloves are completely permeable to these chemicals. Some chemicals can accumulate in certain types of gloves resulting in worsening of the dermatitis.

A much more drastic and often less achievable approach is automation of the working procedure so that manual contact becomes unnecessary. An illustrative example is the introduction of computer-assisted CNS-lathes in which exposure to cutting oil is minimized. An easily neglected part of the treatment is improvement of the work hygiene of the individual worker. It requires a continuous process of education and personal counseling with only long-term results. The contribution of topical steroids, emollients and moisturizers in the spectrum of the treatment is only modest.

Finally, complete eradication of some notorious allergens like glyceryl thioglycolate in permanent wave solution or chromate in cement is mostly out of the range of the individual employer. Such a fundamental step needs a thorough, often years-long debate at the international level between manufacturers and customers, sometimes in close cooperation with international health authorities.

Acknowledgement
I express my gratitude to Dr R.H. Winterkorn, MD for linguistic correction of the manuscript.

Further reading

JG Marks Jr, VA Deleo (eds) (1997) *Contact and occupational dermatology*, 2nd ed, Mosby, St. Louis, USA

JS Taylor (ed) (1994) Occupational dermatoses. *Dermatologic Clinics* 12/3, W.B. Saunders Co, Philadelphia, USA

RM Adams (ed) (1990) *Occupational skin disease*, Saunders Co, Philadelphia, USA

T Menne, HI Maibach (eds) (1994) *Hand eczema*, CRC Press, Boca Raton, USA

RL Rietschel, JF Fowler Jr (eds) (1995) *Fisher's contact dermatitis*, 4th ed, Williams and Wilkins, Baltimore, USA

Clinical aspects: urticaria

Clive E. Grattan

Dermatology Centre, West Norwich Hospital, Norwich NR2 3TU, UK

Definition

Urticaria is characterised by itchy and sometimes painful swellings of the skin or mouth due to transient plasma leakage. It has a spectrum of clinical presentations but is invariably short-lived and resolves without permanent damage. Superficial swellings known as wheals may be white initially due to intense edema with a surrounding red neurogenic flare but become pink as they mature. Wheals may vary in size from a few millimeters to many centimeters across. Deeper swellings called angioedema are often large and poorly defined. They may be pale rather than pink and last longer. Wheals and angioedema can occur in the same individual at the same time. A distinction between them may not be obvious as one may merge with the other. At the very severe end of the spectrum urticaria may occasionally lead to anaphylactic shock and is nearly always a feature of anaphylaxis.

Classification

Urticaria is usually classified by clinical features (Tab. 1) but the groups are not mutually exclusive and some authors prefer a classification based on etiology (Tab. 2). For instance, ordinary and physical urticarias may occur in the same patient but nevertheless have different etiologies. IgE-dependent immediate hypersensitivity may present with contact urticaria, acute ordinary urticaria or anaphylaxis. Angioedema may occur with any pattern of urticaria, including urticarial vasculitis. Hereditary angioedema due to C1 inhibitor deficiency should be considered if angioedema occurs without wheals.

Urticaria must be distinguished from urticarial eruptions in which plasma leakage is only one component of a more persistent pattern of inflammation, as seen in some drug reactions, parasitic infections, papular urticaria from insect bites, erythema multiforme, pemphigoid and polymorphic eruption of pregnancy. Urticaria

Table 1 - Clinical classification of urticaria

Ordinary urticaria	acute (less than 6 weeks of continuous activity)
	chronic (more than 6 weeks of continuous activity)
	episodic (intermittent acute)
Physical urticaria	Symptomatic dermographism
	Heat-induced urticaria
	Exercise-induced anaphylaxis
	Stress-induced (adrenergic) urticaria
	Exercise-induced urticaria
	Cold urticarias
	Aquagenic urticaria
	Solar urticaria
	Vibratory angioedema
	Delayed pressure urticaria
Contact urticaria	Immunological
	Non-immunological
	Uncertain
Urticarial vasculitis	
Angioedema without wheals	Idiopathic
	C 1 esterase inhibitor deficiency
	hereditary (Types 1 and 2)
	acquired (lymphoma, paraprotein and autoimmune-associated)
	Systemic capillary leak syndrome
	Episodic angioedema with eosinophilia
	Secondary to angiotensin converting enzyme inhibitors

pigmentosa is not usually included in classifications of urticaria because the lesions result from locally increased mast cell numbers rather than enhanced cutaneous mast cell releasability.

Clinical features

Urticaria is a common condition. The lifetime prevalence of urticaria may be as high as 10% [1]. It may occur at any age from infancy onwards with a peak incidence in the third decade, with more women being affected than men [2]. About 25% of urticaria patients will have chronic disease at some time [3].

Table 2 - Etiological classification of urticaria

Immunological urticaria	IgE-dependent
	Allergic urticaria
	Immunological contact urticaria
	Autoimmune
	Chronic "idiopathic" urticaria (anti-IgE and anti-FcεRI)
	Complement dependent
	Hereditary angioedema
	Acquired C1 inh deficiency
	Urticarial vasculitis
	Urticarial reactions to blood products
Non-immunological urticaria	Direct mast cell releasing agents
	Opiates
	Curare
	Dextran
	Secondary to pseudoallergens
	Aspirin and nonsteroidal antiinflammatory drugs
	Food additives (e.g. benzoates and azo dyes)
	Natural salicylates
	Angiotensin converting enzyme inhibitor related angioedema
Unknown (idiopathic)	

Ordinary urticaria

This group includes recurrent urticaria where the cause can not be defined clinically by a full history, physical challenge, skin biopsy or conventional blood tests. Many cases remain idiopathic after full evaluation. The wheals range in size from small urticated papules to giant coalescing plaques but, unfortunately, their morphology does not help to define a cause. They are most itchy as they erupt, last no more than 24 h and fade without bruising. About 50% of ordinary urticaria patients also develop angioedema. The urticaria is called acute if attacks continue for less than 6 weeks and chronic if they continue longer. Urticaria lasting for only a few days with prolonged periods of freedom in between acute attacks is known as episodic or acute intermittent. Wheals may erupt every day or less often but chronic urticaria is usually defined for clinical studies as occurring at least twice a week. Many patients note a diurnal pattern to the attacks, often waking with new urticaria in the morning. Women may experience premenstrual exacerbations. They are affected twice as often as men. Measures of the impact of chronic urticaria on the quality of life showed similar scores to ischaemic heart disease for some parameters including social isolation, emotional reactions and energy and worse scores for sleep disruption with the Nottingham Health profile [4], emphasising the disabling effects of the condition. The prognosis of chronic urticaria is worse if associated with angioedema. Fifty percent of patients with wheals alone were clear within 6 months but 50% of those with angioedema as well still had active disease after 5 years [5].

Physical urticaria

This group of disorders is defined by the physical stimulus which provokes urticaria. The underlying etiology remains elusive but passive transfer studies have shown that an immunoglobulin, usually of IgE class, may be involved. Physical urticaria typically presents within 10 min of the stimulus and resolves in less than 2 h, with the exception of delayed pressure urticaria which has a much slower onset and resolution. Angioedema and anaphylaxis may rarely occur with any pattern of physical urticaria. The diagnosis can often be suspected from the history and confirmed in the clinic by simple challenge tests. Diagnostic criteria and guidelines for testing have been produced [6, 7]. Physical urticarias usually arise in adult life and may last for years.

Symptomatic dermographism

Is the commonest form of physical urticaria. Linear wheals arise after stroking the skin. The wheals may coalesce and become so extensive that patients do not recognise they are due to scratching. Most healthy individuals show a dermographic response to firm blunt stroking of the skin but do not experience irritation (simple dermographism). In symptomatic dermographism the whealing threshold is lower and spontaneous irritation triggers a desire to scratch.

Heat-induced urticaria

Cholinergic urticaria is the commonest and best recognised form. It is triggered in affected individuals by a small rise in core temperature through exercise, hot bathing or emotional stress and is often, though not invariably, associated with sweating. The term cholinergic was given because lesions can be reproduced in some patients by intradermal injection of cholinergic agents and inhibited by anticholinergic drugs. It is believed that release of acetylcholine from sympathetic nerve endings supplying eccrine sweat glands leads to mast cell degranulation, perhaps through an intermediate factor which has not been identified. Numerous pale papular wheals surrounded by a pink flare are seen predominantly on the trunk and fade within 2 h. Localised heat urticaria due to direct skin heating is rare and does not appear to involve acetylcholine or noradrenaline release.

Exercise-induced anaphylaxis

Exercise-induced anaphylaxis shows clinical similarities to cholinergic urticaria. The two conditions may represent a clinical spectrum [8]. It has been associated with eating certain foods, such as celery, when it is known as food-dependent, exercise-induced anaphylaxis [9]. Specific IgE to the food can usually be demonstrated by skin-prick testing.

Stress-induced (adrenergic) urticaria

Adrenergic urticaria, characterised by small red macules or papules surrounded by a pale halo, has also been described in a few patients. The lesions, which are brought on by emotional stress, can be reproduced by intradermal noradrenaline and prevented by beta blockade.

Cold urticaria

Several variants of cold urticaria have been described. The commonest form is acquired cold contact urticaria which can be reproduced by melting ice cubes in a plastic bag on the skin or immersion of an arm in cold water. Affected individuals often associate whealing with cold winds or rewarming and are at risk of anaphylaxis with swimming in cold water. Cryoglobulins, cold agglutinins, cold haemolysins and cryofibrinogens may be detected in a few cases but cold urticaria is rare in cryoglobulinaemia. Occasionally, reflex cold urticaria will occur outside the chilled area or on generalised cooling of the body. Cold urticaria may be familial.

Aquagenic urticaria

Is characterised by scattered papular wheals on the trunk surrounded by a wide flare. It is triggered by water contact at any temperature.

Solar urticaria

This rare form of physical urticaria occurs on light-exposed skin within 30 min of exposure to sunlight. It can be demonstrated by phototesting with ultraviolet (280–400 nm) or visible (400–760 nm) light. A light activated serum factor may be involved in its pathogenesis [10].

Vibratory angioedema

Another rare form of urticaria characterised by itchy swellings (often on the limbs) within 10 min of contact with a vibrating stimulus, such as a drill or lawn mower. It may be familial.

Delayed pressure urticaria

May occur alone or in association with chronic idiopathic urticaria. It differs from other forms of physical urticaria by occurring 3–6 h after sustained pressure contact and lasting 24–72 h. The histology resembles a late cutaneous reaction. Systemic symptoms of fever and malaise may occur with attacks and a case of anaphylaxis after pressure testing has been reported [11].

Contact urticaria

Contact urticaria may follow penetration of allergens through the skin or mucous membranes in presensitised individuals or with nonimmunological stimuli. Urticaria tends to be confined to the areas of contact but may become generalised and even progress to anaphylaxis in a few individuals who are highly sensitive to an allergen (e.g. latex). Contact urticaria syndrome may present with nonspecific burning, stinging and itching [12] without typical wheals. This may be a particular problem with cosmetics. Type 1 immediate hypersensitivity has been shown for many foods (e.g. fish, nuts, apples and spices), some drugs (e.g. penicillin G, suxemethonium) and animal products (e.g. epithelia, saliva, insect venoms). Immunological contact urticaria is common in children with atopic eczema but appears to become less so with age. Nonimmunological contact urticaria can be caused by penetration of naturally occurring substances in some plants (e.g. nettlehairs) and animals (e.g. jellyfish) which cause wheals by inducing vasopermeablity or direct mast cell degranulation. The mechanism by which some substances cause contact urticaria is less certain.

Urticarial vasculitis

This uncommon presentation of small vessel vasculitis is often included in classifications of urticaria because it may present with wheals and angioedema that are morphologically similar to chronic ordinary urticaria. However, the primary abnormality is blood vessel damage leading to plasma leakage rather than the transient and reversible capillary leakage seen in other patterns of urticaria. The urticarial lesions sometimes seen in systemic lupus erythematosus usually show leukocytoclastic vasculitis on biopsy even though they resemble chronic urticaria clinically

[13]. The wheals of urticarial vasculitis last days rather than hours. They tend to burn rather than itch. Bruising and induration may be obvious and the patient may show other symptoms and signs of this systemic disorder, such as arthralgia, abdominal pain, malaise and haematuria. The diagnosis can only be established by finding evidence of vasculitis on a skin biopsy. The histological criteria for vasculitis include fibrin deposition, leukocytoclasis, red cell leakage and endothelial cell disruption. The erythrocyte sedimentation rate (ESR) is often raised. Patients with hypocomplementaemia tend to have a more severe clinical course. A syndrome comprising IgM paraproteinaemia, fever, arthralgia and bone pain with urticarial lesions showing neutrophilic or vasculitic histology was first described by Schnitzler [14].

Angioedema without wheals

Most cases of angioedema without wheals are idiopathic. Exclusion of C1 esterase inhibitor deficiency (C1 inh) is essential because the pathogenesis, prognosis and treatment differ from other causes of angioedema. There is a risk of fatal larnygeal oedema without appropriate therapy. C1 inh deficiency may be hereditary or acquired. Type 1 hereditary angioedema (HAE) is due to a quantitative deficiency of C1 inh, whereas the less common Type 2 HAE is due to production of qualitatively nonfunctional C1 inh. C4 levels are reduced during and between attacks with both types and therefore offer an inexpensive screening test. Treatment is aimed at promoting endogenous formation of functional inhibitor with anabolic steroids, stabilising plasmin and covering emergencies with inhibitor replacement therapy. Acquired deficiency of C1 inh due to increased catabolism of C1 inh may occasionally be seen with lymphoma, autoimmune diseases and paraproteinemias. The systemic capillary leak syndrome (Clarkson's syndrome) is characterised by episodic attacks of increased capillary permeability leading to hypotension, fluid retention and generalised angioedema [15]. It is associated with an IgG paraprotein and C4 but C1 inh is normal. Other rare causes of angioedema include episodic angioedema with eosinophilia [16] and the angioedema seen occasionally with angiotensin converting enzyme (ACE) inhibitor therapy [17]. The latter is thought to result from inhibition of kinin degradation by ACE and probably does not involve histamine.

Etiology

As befits such a diverse clinical spectrum there are several potential etiologies for urticaria and a number of associations which have been described but proof of causation is often difficult to obtain.

Allergic urticaria

The binding of allergen to specific IgE on mast cells or basophils causes immuno-logical contact urticaria and many cases of anaphylaxis but is probably an infre-quent explanation for other patterns. As few as 11% of children had food allergy as a cause of their acute urticaria in a recent prospective study [18]. A diagnosis of food allergy is often suspected but hardly ever proven in chronic ordinary urticaria. Atopy is not found more frequently in urticaria than in controls [5].

Autoimmune urticaria

There is a growing body of evidence that many "idiopathic" cases of chronic urticaria may have an autoimmune etiology. Histamine releasing autoantibodies have been found in the sera of 30–50% of chronic urticaria patients in different world centres [19–22]. The majority are directed against the α-subunit of the high affinity IgE (FcϵRIα) receptors on basophils and skin mast cells [23] but about 20% have functional properties of anti-IgE [24]. Anti-FcϵRIα autoantibodies in chronic urticaria belong predominantly to the complement-fixing IgG1 and IgG3 subtypes and there is evidence that histamine releasing activity is linked to their complement fixing properties [21]. Antibodies have also been detected against endothelial cells in chronic urticaria and SLE [25]. Further support for an autoimmune background comes from studies of HLA associations in chronic urticaria which have shown that HLA-DR4 and its associated allele HLA-DQ8 are significantly raised when com-pared with a normal control population [26].

Complement-mediated urticaria

Generation of kinin-like peptides by inappropriate activation of the early complement components causes the angioedema of C1 inh deficiency. C1 inh is consumed in acquired deficiency by being bound to an autoantibody or the uncontrolled activation of C1q by anti-idiotypic antibodies. Deposition of circulating immune complexes in the capillary bed appears to trigger a cascade of events, including local activation of complement, influx of neutrophils and the release of proinflammatory mediators from mast cells in urticarial vasculitis. The urticarial reactions that may accompany transfusions of blood products are also due to immune complex formation.

Nonimmunological urticaria

Certain drugs (e.g. opiates, curare, dextran) can cause mast cell degranulation *in vitro* and *in vivo* by a direct action that does not involve the IgE receptor. Although drug reactions of this type are rare in healthy subjects they may be more severe in chronic urticaria due to enhanced releasability of the mast cells and urticaria pigmentosa due to the increased numbers of mast cells in the skin. Urticaria pigmentosa patients are at increased risk of anaphylaxis after bee or wasp stings due to histamine liberators in the venom [27]. The term pseudoallergy has been applied to the delayed reactions that may be seen in chronic urticaria 15 min to 20 h after ingestion of aspirin, nonsteroidal antiinflammatory drugs, natural salicylates and certain food additives (notably azo dyes and benzoates). It seems likely that the urticaria is due to vasoactive effects of leukotrienes, generated as a consequence of cycloxoygenase inhibition.

Associations

There have been many publications on possible associations between urticaria and infections. Upper respiratory tract infections (usually a viral cold) were noted in 40% of 109 patients presenting with acute urticaria [28]. None of these patients progressed to chronic disease. Links have been suggested between chronic urticaria and dental infections [29], *Candida albicans* and food yeasts [30] and, most recently, *Helicobacter pylori* [31]. It is rare to find parasitic infections as a cause of urticaria in the UK although it may be more prevalent in areas of endemic infection. It is not clear whether chronic infection can promote the onset of autoantibody formation or bring out urticaria by some other mechanism. Patients with established urticaria often note exacerbations with mild viral infections, perhaps mediated by cytokine release. The incidence of thyroid autoimmunity is increased in chronic urticaria [32] and some of these patients will have abnormal thyroid function tests. An epidemiological study of 1155 chronic urticaria patients showed no statistical association with cancer [33] although anecdotal cases have been reported.

Pathogenesis

There are many complex interrelating events which lead to the initiation, development and then resolution of urticaria. Some of these are well understood but the full picture still needs clarification. The mast cell is thought to be central to the pathogenesis of histamine-mediated urticaria by releasing a variety of proinflammatory mediators, cytokines and proteases on degranulation. Inflammatory cells are actively recruited into the dermis from the circulation by adhesion molecule expression. It is likely that histamine releasing autoantibodies in plasma initiate or augment the urticarial reaction. Clinical resolution of the wheal precedes the return of histological normality.

Mast cells

There is lack of agreement on whether the numbers of cutaneous mast cells are increased, the same or decreased in chronic urticaria but it has been shown that the histamine content of skin is increased [34], that chronic urticaria mast cells show enhanced spontaneous and induced histamine release to the experimental secretagogue compound 48/80 [35] and that an uncharacterised histamine releasing factor is present in lesional skin [36] which can be reduced by treatment with corticosteroids [37]. Releasability of skin mast cells to the immunological stimulus, anti-IgE, appears unchanged [38]. Chronic urticaria skin shows an enhanced whealing response to the neuropeptide vasoactive intestinal polypeptide [39] and kallikrein [40]. Mast cells from different tissues show heterogeneity of function, staining characteristics and cytokine profile which might account for the restriction of their response to skin and mucous membranes in urticaria. For instance, cutaneous mast cells release histamine with substance P and morphine but mast cells from other tissues do not [41]. Human cutaneous mast cells produce tumour necrosis factor alpha [42], interleukin-4 [43] and interleukin-8 [44].

Basophils

Chronic urticaria peripheral blood basophils are reduced or absent in the blood of patients with strong histamine releasing activity [45]. Basopenia may therefore be a useful marker for this subgroup of patients. Migration of basophils into the wheals of chronic urticaria may contribute to their persistence by secreting histamine, analogous to the late cutaneous reaction where basophil numbers are increased 14-fold [46]. The observation that they release less histamine than controls when stimulated experimentally with anti-IgE [47] may indicate desensitisation to functional autoantibodies *in vivo*.

Eosinophils

Eosinophils are also present in the late cutaneous response and in the wheals of delayed pressure urticaria. Their role in urticaria is not clear but they may contribute to the increase in vascular permeability by secreting leukotriene C4 and platelet activating factor.

Neutrophils

The neutrophilic histological pattern seen on some urticaria biopsies without an accompanying vasculitis [48] does not appear to be specific for any clinical pattern of urticaria and may represent an acute phase urticarial reaction [49]. The significance of dermal neutrophilia to urticaria pathogenesis remains to be defined.

Investigation of urticaria

Most cases of urticaria do not need investigation if a thorough history and examination have not revealed a cause and the patient has improved spontaneously or responded to antihistamines. Investigations should be driven by clinical suspicion rather than by a desire to "do something".

Blood

Chronic urticaria responding poorly to antihistamines should be investigated with a full blood count and differential for eosinophilia and haematological disorders. The basophil channel on a five-part differential is usually set to count too few cells for accurate estimation in peripheral blood. An ESR will often be raised in urticarial vasculitis but is not specific. Thyroid autoantibodies and thyroid function tests are appropriate if there is a clinical suspicion of thyroid dysfunction. *Helicobacter pylori* serology may be appropriate for chronic urticaria patients with symptoms of gastritis who are unresponsive to conventional antihistamine therapy. C4 is a useful screening test for C1 inh deficiency and should be done in all patients with angioedema alone. Quantitative and functional C1 inhibitor levels are essential in the presence of a low C4 and an indicative history of hereditary angioedema.

Skin tests

Skin-prick tests should be used to support a history of allergic urticaria. Radioallergosorbent tests (RAST) or CAP fluorimmunoassay may be preferred if there is a history of anaphylaxis or to confirm an equivocal skin-prick test result. The autologous serum skin test shows a reasonably high sensitivity (about 70%) and specificity (about 80%) for histamine releasing autoantibodies when defined as a pink wheal at least 1.5 mm larger than an adjacent saline control injection at 30 min [50].

Skin biopsy

A 4 mm punch biopsy from a fresh spontaneous wheal should be performed if vasculitis is suspected clinically. The pattern of histological inflammation in non-vasculitic urticaria does not yield useful information on aetiology but may be of some value in therapy of resistant cases [51].

Challenge tests

The diagnosis of physical urticaria can often be confirmed in the clinic without special resources. The use of double-blinded food additive challenge capsules to define the cause of pseudoallergic reactions is not widespread because the yield of clinically useful information tends to be low but some centres have obtained encouraging results [52].

Therapy

The initial management of urticaria should be directed towards removing or ameliorating the cause when known and minimising exposure to nonspecific aggravating factors, such as overheating, stress, alcohol, opiates (e.g. codeine), aspirin and aspirin-like drugs. Antihistamines are the first line of pharmacological therapy for all types of urticaria except C 1 inh deficiency, ACE inhibitor-related angioedema and episodic angioedema with eosinophilia. The response to antihistamines in urticarial vasculitis and delayed pressure urticaria is often poor and oral corticosteroids or steroid sparing agents may have to be used early. Pseudoallergen exclusion diets may be used for chronic ordinary urticaria patients with an inadequate response to antihistamines and, perhaps, a mast cell stabilising drug. Figure 1 shows a 12-step pyramid strategy for the management of urticaria (excluding C1 inh deficiency).

General measures

The initial assessment should be directed towards identifying and addressing any likely cause of urticaria, such as immediate-type food allergies, contact urticaria, adverse drug reactions, chronic infection (e.g. dental abscess) and thyroid dysfunction. Treatment with thyroxine appeared to be beneficial in a small series of euthyroid chronic urticaria patients with thyroid autoantibodies [53]. Chronic urticaria patients with upper gastrointestinal symptoms and *Helicobacter pylori* infection may respond to eradication therapy [54]. The avoidance of non-specific aggravating factors, such as overheating, stress, high alcohol consumption, and drugs with potential to destabilise urticaria (e.g. aspirin and codeine) should be emphasised early in management. Clearly written information sheets can be very helpful to patients.

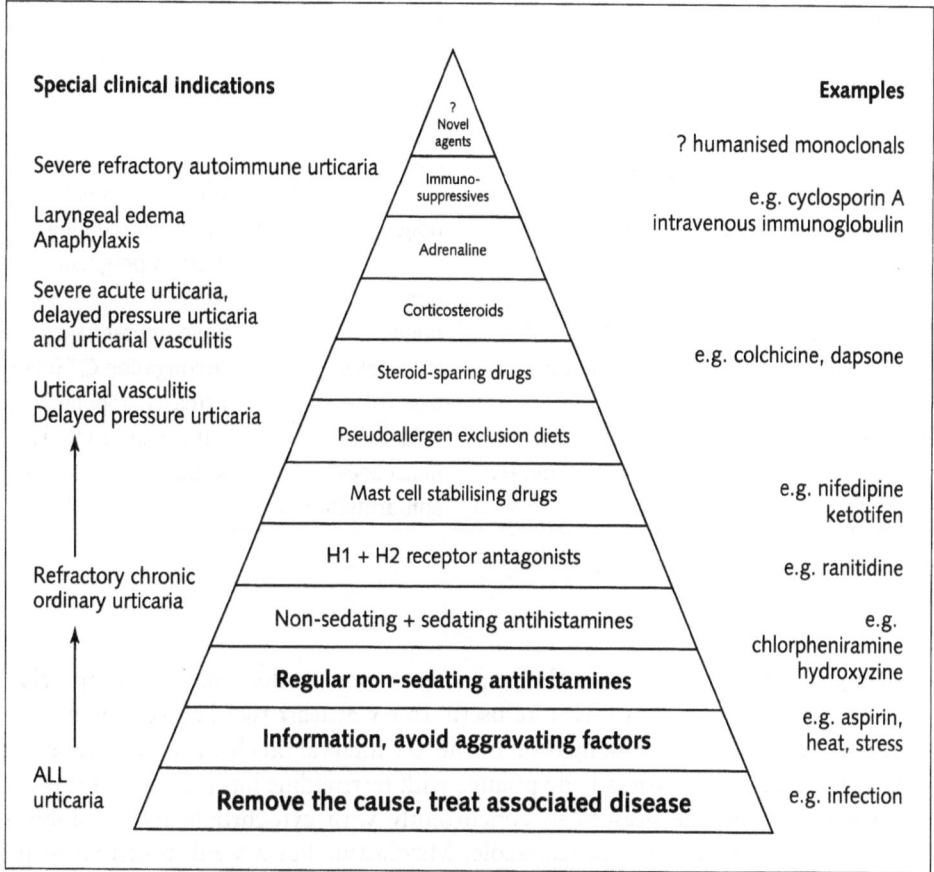

Special clinical indications

Examples

? Novel agents

? humanised monoclonals

Severe refractory autoimmune urticaria

Immuno-suppressives

e.g. cyclosporin A
intravenous immunoglobulin

Laryngeal edema
Anaphylaxis

Adrenaline

Severe acute urticaria,
delayed pressure urticaria
and urticarial vasculitis

Corticosteroids

Steroid-sparing drugs

e.g. colchicine, dapsone

Urticarial vasculitis
Delayed pressure urticaria

Pseudoallergen exclusion diets

Mast cell stabilising drugs

e.g. nifedipine
ketotifen

H1 + H2 receptor antagonists

e.g. ranitidine

Refractory chronic
ordinary urticaria

Non-sedating + sedating antihistamines

e.g.
chlorpheniramine
hydroxyzine

Regular non-sedating antihistamines

Information, avoid aggravating factors

e.g. aspirin,
heat, stress

ALL
urticaria

Remove the cause, treat associated disease

e.g. infection

Figure 1
12-step pyramid management strategy of urticaria

Antihistamines

The efficacy and safety of antihistamines in the therapy of urticaria is undisputed although not all patients respond and, very occasionally, appear to become worse. Seven non-sedating or minimally sedating H1 receptor antagonists are currently available in the UK. Their pharmacological properties are summarised in Table 3. Cetirizine, fexofenadine, loratadine and mizolastine have plasma half-lives of 11–15 h which allows convenient, once-daily dosing. They have a good safety profile with a rapid onset of action and should be considered as a first choice for most urticaria patients. There have been many studies of comparative efficacy of the non-sedating antihistamines from which it appears that terfenadine may be less

Table 3 - Summary of nonsedating antihistamines

Name	Half-life	Dose	Interactions	Special properties
Acrivistine	1.5 h	8 mg t.d.s.	none	short acting, avoid in renal impairment
Cetirizine	11 h	10 mg o.d	none	minimally sedating, avoid in pregnancy
Fexofenadine	11–15 h	180 mg o.d.	none	none
Loratadine	12 h	10 mg o.d.	none	avoid in pregnancy
Mizolastine	13h	10 mg o.d.	imidazoles macrolides	prolongation QT interval increased appetite
Terfenadine	16–32 h	120 mg o.d. OR 60 mg b.d.	imidazoles macrolides anti-arrhythmics	ventricular arrythmias erythema multiforme galactorrhoea

effective than some others but the responses to one product may vary appreciably between patients and it is therefore useful to try at least two before moving on to other groups of drugs. Prolongation of the QT interval on electrocardiographs and the rare occurrence of torsade de pointes with terfenadine limit the use of this drug which should not be prescribed concurrently with cytochrome P450 inhibitors, such as erythromycin and itraconazole. Mizolastine has a weak potential to prolong the QT interval in a few individuals. The active metabolite of terfenadine, fexofenadine, is now licensed for chronic urticaria in the UK and appears to be free of this potential toxicity. Comparative studies with other antihistamines for urticaria are not yet available but initial clinical experience indicates that it compares favourably.

Addition of a sedating antihistamine at night can be helpful when sleep is disturbed by urticaria. A short half-life is desirable so there is little residual sedation the following day. Chlorpheniramine 4 to 12 mg or hydroxyzine 10 to 50 mg are suitable for most patients. The tricyclic antihistamine, doxepin, may be used for its potent antihistaminic properties at doses of 10 to 50 mg and is appropriate for associated depression but anticholinergic effects can be dose limiting and drug interactions need to be considered.

H2-receptor antagonists are not effective on their own for urticaria even though H2 receptors have been demonstrated on cutaneous blood vessels [55]. Urticaria is a rare side-effect of monotherapy. Combinations of H1 and H2 antagonists have been shown in randomised controlled studies to offer additional benefit to the H1 antagonist alone [56, 57] but the overall clinical improvement is often disappointing. Addition of an H2 antagonist, such as ranitidine 150 mg bd, is logical when indigestion or other symptoms of gastritis are present.

Mast cell stabilising drugs

Drugs with theoretical mast cell stabilising properties tend to be disappointing but may be worth trying if combined H1 and H2 antagonists are ineffective. Nifedipine 10 to 20 mg t.d.s. was beneficial when added to existing therapy with H1 and H2 antagonists in a small randomised crossover study [58] and when compared with chlorpheniramine at a dose of 10 mg q.d.s [59]. In view of recent concerns about the safety of short-acting calcium channel blockers in hypertension and ischaemic heart disease it seems prudent to use intermediate or longer acting calcium channel blockers, for which there is currently no clear evidence of increased mortality, in chronic urticaria patients with coexisting hypertension. Ketotifen 1 to 2 mg bd has been used with anecdotal success in chronic urticaria although sedation limits its usefulness. Sodium cromoglycate is not beneficial for urticaria, probably because it is absorbed poorly from the intestine.

Steroid sparing drugs

Steroid sparing drugs, including dapsone, indomethacin, colchicine, azathioprine and hydroxychloroquine have been used in urticarial vasculitis with varying success [60]. The use of dapsone, indomethacin and colchicine in delayed pressure urticaria is disappointing in most patients who may require systemic corticosteroids for control of severe symptoms at the lowest achievable dose. Sulphasalazine has been reported anecdotally for delayed pressure urticaria [61] and corticosteroid dependent chronic idiopathic urticaria [62]. Colchicine has also been used for the neutrophilic pattern of urticaria [51].

Steroids and adrenaline

Oral corticosteroids are often used with benefit for severe acute urticaria, laryngeal angioedema (not due to C1 inh deficiency) and anaphylaxis but have little place in the longterm therapy of chronic ordinary urticaria because of the risk of important unwanted effects, including hypertension, diabetes, peptic ulceration and osteoporosis. Parenteral adrenaline can be life-saving in anaphylaxis (0.3 to 1.0 ml of 1:1000 intramuscularly or subcutaneously) and in laryngeal edema but should be used with caution in hypertension and ischaemic heart disease. It is not recommended for the emergency treatment of angioedema due to C1 inh deficiency.

Immunosuppressive therapies

There is an increasing literature on the use of immunosuppressive therapies in severe chronic autoimmune urticaria refractory to other treatments. Benefit has been suggested in open studies of plasmapheresis [63], intravenous immunoglobulin [64] and cyclosporin A [65]. The value of cyclosporin A has recently been confirmed in a double-blind placebo-controlled randomised study using the Sandimmun (Novartis) formulation at 4 mg/kg/day for 4 weeks with cetirizine (C.E. Grattan, unpublished data).

Novel therapies

Novel therapies for the future may include humanised monoclonal antibodies targeted at IgE or the FcεRI, or peptides based on the α-subunit of FcεRI for neutralisation of circulating autoantibodies in chronic autoimmune urticaria. The development of new antagonists to proinflammatory mediators and their receptors may lead to better therapies.

References

1   Sheldon JM, Matthews KP, Lovell RG (1954) The vexing urticaria problem. Present concepts of aetiology and management. *J Allergy* 25: 525–560

2   Humphreys F, Hunter JAA (1998) The characteristics of urticaria in 390 patients. *Br J Dermatol* 138: 635–638

3   Greaves MW (1995) Chronic urticaria. *N Engl J Med* 332: 1767–1772

4   O'Donnell BF, Lawlor F, Simpson J, Morgan M, Greaves MW (1997) The impact of chronic urticaria on the quality of life. *Br J Dermatol* 136: 197–201

5   Champion RH, Roberts SOB, Carpenter RG, Roger JH (1969) Urticaria and angio-oedema: a review of 554 patients. *Br J Dermatol* 81: 588–597

6   Kobza Black A, Lawlor F, Greaves MW (1996) Concensus meeting on the definition of physical urticarias and urticarial vasculitis. *Clin Exp Dermatol* 21: 424–426

7   Kontou-Fili K, Borici-Mazi R, Kapp A, Matjevic LJ, Mitchel FB (1996) Physical urticaria: classification and diagnostic guidelines. *Allergy* 52: 504–513

8   Volcheck GW, James TC (1997) Exercise-induced urticaria and anaphylaxis. *Mayo Clin Proc* 72: 140–147

9   Kidd JM, Cohen SH, Sosman AJ, Fink JN (1983) Food-dependent exercise-induced anaphylaxis. *J Allergy Clin Immunol* 71: 407–411

10 Horio T, Minami K (1977) Solar urticaria: photoallergen in a patient's serum. *Arch Dermatol* 113: 157–160

11 Mijailovic BB, Karadaglic DjM, Ninkovic MP, Mladenovic TM, Zecevic RD, Pavlovic

MD (1997) Bullous delayed pressure urticaria; pressure testing may produce a systemic reaction. *Br J Dermatol* 136: 434–436.

12   von Krogh G, Maibach HI (1981) The contact urticaria syndrome – an updated review. *J Am Acad Dermatol* 5: 328–342

13   O'Laughlin S, Schroeter AL, Jordon RE (1978) Chronic urticaria-like lesions in systemic lupus erythematosus. *Arch Dermatol* 114: 879–883

14   Schnitzler L, Schubert B, Boasson M, Gardais J, Tourmen A (1974) Urticaire chronique, lésions osseuses, macroglobuliémie IgM: maladie de Waldenström? *Bull Soc Franc Dermatologic Syphilgraphie* 81: 363

15   Gleich G J, Schroeter AL, Marcoux JP, Sachs MI, O'Connell EJ, Kohler PF (1984) Episodic angioedema associated with eosinophilia. *N Engl J Med* 310: 1621–1626

16   Clarkson B, Thompson D, Horwith M, Luckey EH (1960) Cyclical edema and shock due to increased capillary permeability. *Am J Med* 29: 193–216

17   Sabroe RA, Kobza Black A (1997) Angiotensin-converting enzyme (ACE) inhibitors and angio-oedema. *Br J Dermatol* 136: 153–158

18   Mortureux P, Léauté-Labrèze C, Legrain-Lifermann V, Lamireau T, Sarlangue J, Taïeb A (1998) Acute urticaria in infancy and early childhood: a prospective study. *Arch Dermatol* 134: 319–323

19   Niimi N, Francis DM, Kermani F, O'Donnell BF, Hide M, Kobza Black A, Winkelmann RK, Greaves MW, Barr RM (1996) Dermal mast cell activation by autoantibodies against the high affinity IgE receptor in chronic urticaria. *J Invest Dermatol* 106: 1001–1006

20   Zweiman B, Valenzano M, Atkins PC, Tanus T, Getsy JA (1996) Characteristics of histamine-releasing activity in the sera of patients with chronic urticaria. *J Allergy Clin Immunol* 98: 89–98

21   Fiebiger E, Hammerschmid F, Stingl G, Maurer D (1998) Anti-FcεRIα autoantibodies in autoimmune-mediated disorders: identification of structure-function relationship. *J Clin Invest* 101: 243–251

22   Ferrer M, Kinét J-P, Kaplan AP (1998) Comparative studies of functional and binding assays for IgG anti-FcεRIα (α-subunit) in chronic urticaria. *J Allergy Clin Immunol* 101: 672–6

23   Hide M, Francis DM, Grattan CEH, Hakimi J, Kochan JP, Greaves MW (1993) Autoantibodies against the high-affinity IgE receptor as a cause of histamine release in chronic urticaria. *N Engl J Med* 328: 1599–1604

24   Grattan CEH, Francis DM, Hide M, Greaves MW (1991) Detection of ciculating histamine releasing autoantibodies with functional properties of anti-IgE in chronic urticaria. *Clin Exp Allergy* 21: 695–704

25   Grattan CEH, D'Cruz DP, Francis DM (1995) Antiendothelial cell antibodies in chronic urticaria. *Clin Exp Rheumatol* 13: 272–273

26   O'Donnell BF, O'Neill CM, Francis DM, Niimi N, Barr RM, Barlow RJ, Kobza Black A, Welsh KI, Greaves MW (1999) HLA Class II associations in chronic idiopathic urticaria. *Br J Dermatol* 140: 853–858

27   Henz BM, Zuberbier T (1998) Causes of urticaria. In: BM Henz, T Zuberbier, J Grabbe, E Monroe (eds): *Urticaria: clinical, diagnostic and therapeutic aspects*. Springer-Verlag, Berlin

28   Zuberbier T, Iffländer J, Semler C, Henz BM (1996) Acute urticaria: clinical aspects and therapeutic responsiveness. *Act Derm Venereol (Stockh)* 76: 296–298

29   Resch CA, Evans RR (1958) Chronic urticaria and dental infection. Cleveland Clinic Quarterly 25: 147–150

30   James J, Warin R (1971) An assessment of the role of *Candida albicans* and food yeasts in chronic urticaria. *Br J Dermatol* 84: 227–237

31   Tebbe B, Geilen CC, Jörg-Dieter Schulzke, Bojarski C, Radenhausen M, Orfanos CE (1996) Helicobacter pylori infection and chronic urticaria. *J Am Acad Dermatol* 34: 685–686

32   Leznoff A, Sussman GL (1989) Syndrome of idiopathic chronic urticaria and angioedema with thyroid autoimmunity: a study in 90 patients. *J Allergy Clin Immunol* 84: 66–71

33   Lindelöf B, Sigurgeirsson B, Wahlgren CF, Eklund G (1990) Chronic urticaria and cancer: an epidemiological study of 1155 patients. *Br J Dermatol* 123: 453–456

34   Phanuphak P, Schocket AL, Arroyave CM, Kohler PF (1980) Skin histamine in chronic urticaria. *J Allergy Clin Immunol* 65: 371–375

35   Bédard PM, Brunet C, Pelletier G, Hébert J (1986) Increased compound 48/40 induced local histamine release from lesional skin of patients with chronic urticaria. *J Allergy Clin Immunol* 78: 1121–1125

36   Claveau J, Lavoie A, Brunet C, Bédard P-M, Hébert J (1993) Chronic idiopathic urticaria: possible contribution of histamine releasing factor to pathogenesis. *J Allergy Clin Immunol* 92: 132–137

37   Paradis L, Lavoie A, Brunet C, Bédard P-M, Hébert J (1996) Effects of systemic corticosteroids on cutaneous histamine secretion and histamine-releasing factor in patients with chronic idiopathic urticaria. *Clin Exp Allergy* 26: 815–820

38   Zuberbier T, Schwarz S, Hartmann K, Pfrommer C, Czarnetzki BM (1996) Histamine releasability of basophils and skin mast cells in chronic urticaria. *Allergy* 51: 24–28

39   Juhlin L, Michaëlsson G (1969) Cutaneous reactions to kallikrein, bradykinin and histamine in healthy subjects and in patients with urticaria. *Acta Derm Venereol* 49: 26–36

40   Smith CH, Atkinson B, Morris RW, Hayes N, Foreman JC, Lee TH (1992) Cutaneous responses to vasoactive intestinal polypeptide in chronic urticaria. *Lancet* 339: 91–93

41   Lowman MA, Rees PH, Benyon RC, Church MK (1988) Human mast cell heterogeneity: histamine release from mast cells dispersed from skin, ling, adenoids, tonsils, and colon in response to IgE-dependent and nonimmunologic stimuli. *J Allergy Clin Immunol* 81: 590–597

42   Walsh LJ, Tinchiere G, Waldorf HA, Whitaker D, Murphy GF (1991) Human dermal mast cells contain and release tumour necrosis factor-α, which induces endothelial leukocyte adhesion molecule 1. *Proc Natl Acad Sci* 88: 4220–4224

43   Bradding P, Feather IH, Howarth PH, Mueller R, Roberts JA, Britten K, Bews JPA, Hunt

TC, Okayama Y, Heusser CH et al (1992) Interleukin 4 is localised to and released by human mast cells. *J Exp Med* 176: 1381–1386

44 Möller A, Lippert U, Lessman D, Kolde G, Hamann K, Welker P, Schadendorf D, Rosenbach T, Luger T, Czarnetski BM (1993) Human mast cells produce IL-8. *J Immunol* 151: 3261–3266

45 Grattan CEH, Walpole D, Francis DM, Niimi N, Dootson G, Edler S, Corbett MF, Barr RM (1997) Flow cytometric analysis of basophil numbers in chronic urticaria: basopenia is related to serum histamine releasing activity. *Clin Exp Allergy* 27: 1417–1424

46 Charlesworth EN, Hood AF, Soter NA, Kagey-Sobotka A, Norman PS, Lichtenstien LM (1989) Cutaneous late-phase response to allergen. *J Clin Invest* 83: 1519–1526

47 Greaves MW, Plummer VM, McLaughlan P, Stanworth DR (1974) Serum and cell bound IgE in chronic urticaria. *Clin Allergy* 4: 265–271

48 Peters MS, Winkelmann RK (1985) Neutrophilic urticaria. *Br J Dermatol* 113: 25–30

49 Toppe E, Haas N, Henz BM (1998) Neutrophilic urticaria: clinical features, histological changes and possible mechanisms. *Br J Dermatol* 138: 248–253

50 Sabroe RA, Grattan CEH, Francis DM, Barr RM, Kobza Black A, Greaves MW (1999) The autologous serum skin test: a screening test for autoantibodies in chronic idiopathic urticaria. *Br J Dermatol* 140: 446–452

51 Zavadak D, Tharp MD (1995) Chronic urticaria as a manifestation of the late phase reaction In: EN Charlesworth (ed): *Immunology and allergy clinics of North America. Urticaria*, WB Saunders, Philadelphia, 745–760

52 Zuberbier T, Chantraine-Hess S, Harmann K, Czarnetski BM (1995) Pseudoallergen-free diet in the treatment of chronic urticaria. A prospective study. *Acta Derm Venereol (Stockh)* 75: 484–487

53 Rumbyrt JS, Katz JL, Schocket AL (1995) Resolution of chronic urticaria in patients with thyroid autoimmunity. *J Allergy Clin Immunol* 96: 901–905

54 di Campli C, Gasbarrini A, Nucera E, Franceschi F, Ojetti V, Torre ES, Schiavino D, Pola P, Patriarca G, Gasbarrini G (1998) Beneficial effects of *Helicobacter pylori* eradication on idiopathic chronic urticaria. *Digestive Dis Sci* 43: 1226–1229

55 Robertson I, Greaves MW (1978) Responses of human skin blood vessels to synthetic histamine analogues. *Br J Clin Pharmacol* 5: 319–322

56 Bleehen SS, Thomas SE, Greaves MW, Newton J, Kennedy CTC, Hindley F, Marks R, Hazell M, Rowell NR, Fairiss GM et al (1987) Cimetidine and chlorpheniramine in the treatment of chronic idiopathic urticaria: a multi-centre randomised double-blind study. *Br J Dermatol* 117: 81–88

57 Paul E, Bödeker RH (1986) Treatment of chronic urticaria with terfenadine and ranitidine: a randomised double-blind study in 45 patients. *Eur J Clin Pharmacol* 31: 277–280

58 Bressler RB, Sowell K, Huston DP (1989) Therapy of chronic idiopathic urticaria with nifedipine: demonstration of beneficial effect in a double-blinded, placebo controlled, crossover trial. *J Allergy Clin Immunol* 83: 756–763

59   Liu H-N, Pan L-M, Hwang S-C, Chi T-L (1990) Nifedipine for the treatment of chronic urticaria: a double-blind cross-over study. *J Dermatol Treatment* 1: 187–189

60   National Prescribing Centre (1998) Safety of calcium-channel blockers. *MeReC Bulletin* 9: 13–16

61   Engler RJM, Squire E, Benson P (1995) Chronic sulfasalazine therapy in the treatment of delayed pressure urticaria and angioedema. *Ann Allergy Asthma Immunol* 74: 155–159

62   Jaffer AM (1991) Sulfasalazine in the treatment of corticosteroid-dependent chronic idiopathic urticaria. *J Allergy Clin Immunol* 88: 964–965

63   Grattan CEH, Francis DM, Slater NGP, Barlow RJ, Greaves MW (1992) Plasmapheresis for severe, unremitting, chronic urticaria. *Lancet* 339: 1078–1080

64   O'Donnell BF, Barr RM, Kobza Black A, Francis DM, Kermani F, Niimi N, Barlow RJ, Winkelmann RK, Greaves MW (1998) Intravenous immunoglobulin in autoimmune chronic urticaria. *Br J Dermatol* 138: 101–106

65   Toubi E, Blant A, Kessel A, Golan TD (1997) Low-dose cyclosporin A in the treatment of severe chronic idiopathic urticaria. *Allergy* 52: 312–316

Clinical aspects: drug allergy

Hans F. Merk

Department of Dermatology and Allergology, University Hospital, RWTH Aachen, Pauwelsstr. 30, 52074 Aachen, Germany

Introduction

Drug allergy can be the cause of many inflammatory reactions involving the liver, lung, kidney, GIT and blood cell formation leading to such diseases as agranulocytosis [1–4]. However it is the skin, which is frequently involved in these reactions and relatively overtly, that is considered to be a signalling organ for these reactions [5]. The clinical signs and symptoms of these reactions are very heterogenous and include all different types of allergic reactions such as immediate type of reactions including anaphylaxis, type II and III reactions such as purpura and hypersensitivity vasculitis as well as delayed type reactions such as allergic contact dermatitis, maculo-papular rashes, fixed drug eruptions, photoallergy, and bullous drug reactions including the toxic epidermal necrolysis [6]. The skin is also quite frequently used as test organ for diagnostic purposes by using the prick test, intracutaneous test or patch test including photo-patch test in the case of photosensitivity reactions [7]. It has been shown that the skin has not only the full armentarium to develop sensitivity such as antigen presenting cells, lymphocytes and allergic reactions modifying cytokine producing keratinocytes but also those enzymes which are able to metabolize small molecular weight drugs such as cytochrome P450 isoenzymes or transferases [8]. Quite recently it has been found that the skin as an extrahepatic drug-metabolizing organ can convert the anticonvulsant carbamazepine which is quite often involved in the anticonvulsant hypersensitivity syndrome to a highly reactive protein binding derivative [9]. Therefore this chapter will be focused on skin drug reactions.

Epidemiology of cutaneous drug allergy

Idiosyncratic reactions (type B) are a major complication of drug therapy, because they are related to both the drug and to individual factors in the host in opposite to type A which is dependent alone on the pharmacological characteristics of the drug. About 25% to 30% of type B reactions are estimated to be allergic drug reactions.

Immunology and Drug Therapy of Allergic Skin Diseases, edited by C.A.F.M. Bruijnzeel-Koomen and E.F. Knol

The best available data about the frequency of drug allergic reactions exist for penicillin, although they do not permit exact conclusions [10]. They are reported to occur in 0.7% to 8% of treatments in different studies including anaphylaxis in 0.004% to 0.015% of all treatments. Fatality from penicillin by anaphylaxis happens about once in every 50 000 to 100 000 treatments leading to 400–800 deaths per year [11]. Population based retrospective studies were performed in Germany and France in severe bullous drug reactions. They revealed an incidence of this type of drug allergic reactions of 1.2 per 1 million inhibitants per year in France and 0.93 in Germany, respectively. There is an about tenfold risk for these reactions in HIV-infected patients [12].

Immediate type reactions

Symptoms of immediate allergic reactions are urticaria, angioedema, rhinitis, asthma, and anaphylaxis. Mast cells and basophils play a central role in the pathogenesis of these disorders. These cells possess the high affinity receptor for IgE (FcεRI). A particular drug or its metabolites which bind to at least two IgE molecules can cause bridging which leads to a release of inflammatory mediators such as histamine, derivatives of arachidonic acid, and mast cell-derived chemotactic factors for neutrophils and eosinophils; also proteases including tryptase or chymase are activated. Drugs such as β-lactam antibiotics, pyrazolones, sulfonamides and heterogenous proteins including protamin, chymopapain, streptokinase, and trasylol are known to be a major cause of these reactions [13, 14]. Protamin is a protein drug capable of causing anaphylactic reactions, which is used in insulin preparations and cardiovascular surgery with heparin. Antibodies of the IgE class can be detected against protamin, but their prognostic value in the prediction of a risk for anaphylactic reactions in cardiovascular surgery is questionable [15]. Patients who have diabetes, or who have had vasectomies are most susceptible to these reactions. It is unclear yet if low molecular weight substances are predisposing factors because of their chemical properties or because they are frequently prescribed. At least some drugs, such as digoxin are known to be rare causes of allergic reactions, even though they are frequently prescribed. Other examples of those drugs are listed in Table 1.

Pseudoallergic reactions

Pseudoallergic reactions are able to cause clinical features identical to those in allergic reactions, which are also based on the release of inflammatory mediators from mast cells and basophils, but which are IgE-independent. A well-known phenomenon is the pseudoallergic reactions to nonsteroidal antiphlogistic drugs (NSAID). Patients with this syndrome have a high risk of developing symptoms such as

Table 1 - Drugs which are often (A) or very rarely (B) the cause of drug allergy

A	B
β-lactam antibiotics	adrenergic drugs
pyrazolones	androgens
sulfonamides	antihistamines (except topical)
anticonvulsants	atropine
suxamethonium	benzodiazepine
heterogenous proteins	digoxin-digitoxin
• protamin	sodium cromoglycate
• chymopapain	ganglia blockers
• streptokinase	nystatin
• trasylol	estrogens
	thyroid hormones
	spironolactone
	tetracycline
	theophylline
	dicoumarol

urticaria, angioedema, asthma, or even anaphylactic shock on treatment with these substances. The observation that chemically unrelated drugs can cause these reactions, makes it unlikely that these are allergic reactions [16]. New findings with regard to asthma reactions to NSAID suggested that patients with this type of reaction have an increased formation of leukotrienes in contrast with prostaglandin D2 and in these patients these reactions to NSAID are prevented after pretreatment with leukotriene receptor antagonists [17, 18]. β-receptor blockers are also capable of augmenting acute allergic (especially anaphylactic) reactions, because activation of β-receptors inhibits the release of inflammatory mediators from mast cells. These effects may be fatal in the treatment of anaphylactic shock with epinephrine, because paradoxical effects of epinephrine may occur due to blocking of β-receptors [14]. One should be aware of this fact when taking care of patients who are sensitized to food products, spices, or venoms, but also of patients undergoing immunotherapy. Captopril and enalapril, which are inhibitors of the angiotensin converting enzyme (ACE), are known to produce pseudoallergic angioedema in one of 3000 patients. They show not only inhibition of the angiotensin system, but also inactivation of bradykinin and substance P, which are both important mediators of inflammation [19]. Because these reactions are based on the pharmacological properties of the drugs, one substance cannot be substituted by the other. However, not all of the captopril-dependent skin reactions are based on the inhibition of the angiotensin system. For example, pemphiguslike reactions can be provoked by drug with free SH-binding sites. For this reason, patients with pemphigus-like skin lesions can be treated without risk with enalapril, which does not bear free SH groups. These examples show that an exact understanding of the underlying pathogenesis is desirable in order to draw the right conclusion for the continuing treatment of patients who have experienced unexpected drug induced reactions.

Purpura and hypersensitivity vasculitis

Allergic reactions of the late type can show various manifestations of the skin. In most instances they develop several hours after the intake of the particular drug, eg, a purpura, or different signs and symptoms of vasculitis. The underlying pathogenesis is based on the formation of complexes between antibodies and antigens leading to the activation of complement or immune complex binding to cells with cytotoxic effects [3]. The so-called IgE-dependent late phase reaction can be interposed between this group and allergic immediate reactions. Recently, IgE-dependent late phase reactions have gained particular interest, because they are characterized by symptoms similar to allergic immediate reactions such as urticaria, anaphylactic shock, or asthma. Studies on the pathogenesis of IgE-dependent late phase reactions in allergic rhinitis and allergic skin reactions have shown that these symptoms are caused mainly by basophils instead of mast cells: All mediators of inflammation were found except for prostaglandin D2, which is a mast cell product [20]. Manifestations of late type reactions are allergic vasculitis, purpura, pigmentosa, erythema exsudativum multiforme, and serum sickness. The causative agents of these disorders are found only in 10% of all cases by skin tests or *in vitro* assays. Late type reactions due to penicillins were found to depend not only on binding to antibodies of the IgE, but also of the IgG and IgM class. Erythema nodosum is characterized by a profound vasculitis, which is sometimes caused by drugs such as sulfonamides, pyrazolones, and contraceptives [21].

Allergic contact dermatitis and systemic contact dermatitis

Allergic contact dermatitis is the prototype of a classical delayed type allergic reaction. Allergic contact dermatitis was, for example, frequently observed after local application of tromanthadine in the treatment of labial herpes. However even glucocorticoids can occasionally cause allergic contact dermatitis. The method of application seems to be critical for the development of contact dermatitis. Benzoyl peroxide, for example, causes sensitization in 40% of patients with crural ulcers, which therefore is a contraindication for treatment with this substance. In the treatment of acne, however, sensitizations to benzoyl peroxide occur very rarely. Eczematous or exanthematous widespread dermatitis from internal exposure to an allergen to which the patient has previously become sensitized *via* the skin or those allergens which may produce systemic effects after being absorbed by percutantaneous penetration or inhalation are termed systemic contact dermatitis [22]. There are many examples for such reactions including sulphonamides, corticoids, doxepin, mercury, thiomersal, clindamycin, azole derivatives, propylene glycol, salbutamol and other β-receptor antagonists or agonists.

Macular-papular rash

Another manifestation of a delayed allergic reaction is the classical morbilliform exanthems, because those patients react quite often to the culprit drug in the patch test and sensitized peripheral lymphocytes can be shown *in vitro* in affected patients. The signs and symptoms of this disease vary, and may include maculae, papules and even blisters. Sometimes it may appear with the clinical picture of an acute generalized exanthematous pustulosis, which occurs with drugs such as anticonvulsants, tetracycline and erythromycine, phenobarbital, sulfonamids and even retinoids [2]. There is a variety of drugs which can cause morbilliform exanthema, especially β-lactam antibiotics and antiepileptics that seem to be a major cause. Especially in these cases the patch test is quite helpful for finding the diagnosis including the culprit drug. Erythema exsudativum multiforme with well marked single lesions is an episodic, self-limited disorder of the skin, which is most often a clinical symptom of a hypersensitivity reaction to herpes viruses [23].

Fixed drug eruption

The fixed drug eruption is a disorder whose pathogenesis is not yet clearly understood. In most instances, there is a characteristic nummular erythema frequently showing a central blister. The predilection areas are the external genitals and distal parts of the body including fingers, toes and elbows. A more severe variant of fixed drug rection is the generalized bullous fixed eruption which is sometimes mixed up with other bullous drug reactions such as TEN [24]. In some patients, the development of multiple fixed drug eruptions can be observed especially after repeated challenges. Interestingly, the original presentation of Lyell's disease includes these cases. One study indicated that epidermal cells in the affected areas exhibit a pathological regulation mechanism for the expression of adhesion molecules [25]. It is typical for this type of drug allergy that one can sometimes induce this reaction by the topical application of the drug at the site where it occurred. Drugs which are most often involved are barbiturates, phenazones, sulfonamides, phenytoin and trimethoprim [24].

Bullous drug eruption

Bullous drug reactions are the most dangerous reactions for the patient beside IgE-mediated anaphylactic shock. Examples are the fixed drug eruption including multiple fixed drug eruption syndrome, toxic epidermal necrolysis (TEN), Stevens-Johnson syndrome (SJS) and the overlap TEN/SJS (Tab. 2) [26, 27]. Stevens-Johnson syndrome, sometimes characterized as a major variant of erythema multiforme, is much more frequently drug-induced with a primary manifestation at mucous membranes and involves normally less than 10% of the skin. TEN, an extreme form of a bullous drug reaction, is characterized by a sudden epidermal destruction, resulting in loss of the skin. Morphologically, the skin lesions show extensive cell death in the basal and the Malpighian layer especially close to lymphocyte infiltration. Immunohistochemical analysis has identified CD8+ T cells as the predominant epidermal T cell subset in drug induced bullous eruptions to β-lactam antibiotics [28]. These CD8+ T lymphocytes produced a TH1-like cytokine pattern and were cytotoxic under *in vitro* conditions against autologous B cells upon stimulation through the T cell receptor and against epidermal keratinocytes in lectin-induced cytotoxicity assays [29]. Further information on recent studies suggests that the cell destruction in TEN and TEN/SJS may be a result of apoptosis, which will be discussed later in the section of the role of T lymphocyte assays [30].

Photosensitivity reactions

Distinct forms of allergic reactions are those caused by interactions of a drug or its particular metabolite with UV radiation, a clinical feature which is characterized by manifestation of the skin exposed to light [31]. Especially ultraviolet A light (UVA) radiation (mainly produced in tanning booths), is capable of producing severe skin reactions in combination with particular drugs which may even lead to erythroderma. Even commonly applied drugs such as nonsteroidal antiphlogistics are able to induce these reactions. Drugs that produce cutaneous photoxic responses must efficiently absorb radiant energy whose specific wavelength is dependent on the physicochemical properties of the photoxic agent and lies in most instances within the ultraviolet A range. On the other hand, only a few drugs, such as sulfanilamide are known to be phototoxic in connection with ultraviolet B light (UVB) radiation [32]. Beside photoxic reactions photosensitivity reactions may be produced by at least two other mechanisms. First, they may exacerbate diseases which are known to develop photosensitivity such as hydrochlorothiazide-induced subacute cutaneous lupus erythematodes or isoniazid induced pellagra. The second possibility is photoallergy, which involves an immunologically mediated response to the drug. The differentiation between pho-

Table 2 - Differential Diagnose of EMmajor/SJS/TEN [26, 27]

- EM major: mucosal erosions plus skin lesions characteristic in their pattern (targets without blisters) and distribution (symmetrical and mainly acral)
- SJS: Mucosal erosions plus widespread purpuric cutaneous macules often confluent with positive Nikolsky's sign and epidermal detachment
- SJS-TEN: widespread purpuric macules and epidermal detachment between 10%–30%
- TEN: widespread purpuric macules and epidermal detachment above 30% or
- widespread necrolysis (more than 10% detachment) without any discrete lesions

---

toxic and photoallergic reactions is often difficult because many similarities with regard to the clinical and histologic features exist. However the impact of photoallergic reactions for the patient is more severe because there is not only the persistent sensitization, but also the risk to develop a persistent light hypersensitivity with the clinical signs and symptoms of actinic reticuloid [31]. Drugs with phototoxic properties include coal tar, nalidixic acid and several quinones, furosemide, tetracyclines, phenothiazines, amiodarone, NSAIDS or thiazide diuretics, quinidine, quinine, sulfonylurea antidiabetic agents, hematoporphyrin derivatives, vinblastine, dacarbazine, 5-fluorouracil and sodium valproate [31, 32]. A further reason for the difficulty to separate phototoxic reactions from photoallergic one is that some drugs can produce both types of reactions such as phenothiazine or tetracyclines. Further drugs are hydrochlorothiazide, chlorothiazide, tolbutamide, chlorpropamide, griseofulvin, sulfonamides, calcium cyclamate, piroxicam, diphenhydramine, antimalarias, quinidine. Topical agents especially include sunscreen filters such as PABA, benzophenons, digalloyl triolate, dibenzoylmethanes and cinnamates [33]. The diagnosis of phototoxic or photoallergic compounds can be rather difficult. One example is dacarbazine. Under clinical conditions a phototoxic reaction was observed however it was not possible to reproduce this reaction by a photopatch test. Even after oral challenge an increase of photoxicity was not found unless the test was performed 4 h after the challenge, suggesting that a metabolite of dacarbazine is responsible [31]. These examples demonstrate that carefully performed preclinical tests must be designed especially in the case of compounds which are unstable when they are exposed to light or UV-light. The evaluation of appropiate *in vitro* or animal tests such as the lymph node assay are in progress [34].

A photodynamic effect is a phototoxic reaction with activated oxygen metabolites involved. This pathogenetic effect could be shown for tetracycline and its derivatives, a phenomenon which is used today as a therapeutic approach with porphyrins.

Drug-induced autoimmune diseases

There are several hypotheses on the pathogenesis of drug-induced autoimmune diseases which show a great variety of skin manifestations: a drug or its metabolites have a direct pharmacological effect on the immune systems; the drug or its metabolites bind to DNA or to self-proteins which renders them immunogenic; genetic factors such as polymorphisms in drug metabolism or immunological factors (MHC antigens) lead to different reactions in the individual patient [35]. Drugs are capable of producing lupus erythematosus (LE), pemphigus or pemphigoid-like skin disorders. Pemphigus-like skin reactions are primarily caused by drugs bearing a free sulfhydryl (SH) group, whereas LE-like disorders develop frequently after treatment with isoniazid, hydralazine, and procaine [37]. Examples of pemphigus-inducing drugs are d-penicillamine and captopril. These SH-group-bearing drugs can induce acantholysis of keratinocytes even under *in vitro* conditions. There is a case report of one patient who developed drug-induced pemphigus due to captopril as well as d-penicillamine [35]. Patients who slowly metabolize isoniazid, hydralazine, or procaine by acetylation are highly susceptible to drug-induced SLE-like autoimmune diseases [36]. Antibodies against histones were found to be pathogenetic factors in LE-like reactions, which were directed specifically against the H2A-H2B-histone complex when procainamide was the causative agent [37]. The IgG/IgM-antibody ratio is critical for the development of symptoms such as fever, myalgia, arthralgia, pleuritis, pericarditis, and skin reactions which manifest as increased light sensitivity. Less frequently observed are alterations of the central nervous system, kidneys, or mucous membranes. Recently the important role of Fas-FasL dependent apoptotic reactions have been shown in Fas-deficient C57BL/6-lpr/lpr mice [38]. Injection of as few as 5000 chromatin-reactive T cells into naïve, syngeniec mice induced a rapid IgM anti-denatured DNA-response, while injection of at least 100-fold greater number of activated T cells was required for induction of IgG anti-chromatin Abs, suggesting that small numbers of autoreactive T cells can be homeostatically controlled.

In vivo testing for adverse drug reactions

Skin tests are employed in order to reproduce similar clinical signs and symptoms on challenge with the suspected drug. If the response is similar to that observed on primary exposure, such as the triple response of erythema, itch and wheal in a skin test, it is concluded that an allergic reaction has occurred. Under immunological considerations, this is of course far from a proof. Therefore, it is difficult, if not impossible, to separate allergic from pseudoallergic reactions by these tests. Skin-prick test, intradermal test and patch test including photo patch test are performed as described elsewhere in this volume. Special placebo-controlled test protocols are suggested for side-effects to local anesthetics [39–41].

If the need of elucidating the cause of a certain drug reaction exceeds the risk of evoking a severe test reaction, oral challenge testing is justified. Normally one starts with the informed consent of the patient and with 1% or less of the therapeutic dose, followed by two to ten-fold increments at 30 to 60 min intervals. This test should be performed only by experts with emergency equipment at hand. Although this test is the "gold standard", it does not permit differentiation between allergic and pseudoallergic reactions. Newer approches try to combine this test with *in vitro* data such as the increase of cytokines including IL-5, IL-1, IL-6, IFNγ or mast cell mediators, e.g. histamine and tryptase or those of eosinophiles such as ECP or the expression of early markers on T lymphocytes [42] in order to diagnose not just on the basis of clinical observations. However these methods have not yet been evaluated.

In vitro testing for adverse drug reactions and pathophysiology of drug allergy

In vitro tests can be separated for practical reasons into serologic and cellular tests. In serologic tests specific IgE is measured most often by the radio or enzyme allergosorbent test or the immunoblot. The first type of assay is well evaluated only for allergic reactions to penicilloyl derivatives of β-lactam-antibiotics. The immunoblot is especially used if allergic reactions occurred to proteins or peptides such as trasylol.

Cellular tests include the incubation of the suspected culprit drug with basophiles and to measure the release of histamine and leukotriene-derivatives. The basophiles may be preincubated with interleukin 3 – known as CAST-Elisa – to make them more vulnerable to IgE dependent mediator release. Especially most severe anaphylactic reactions may be diagnosed by this test which will principally detect allergic as well as some pseudoallergic reactions, however evaluations have not been performed or published yet [43].

Lymphocyte transformation test (LTT)

Another approach is the LTT. T lymphocytes play a key role in the pathogenesis of allergic drug reactions because they are not only involved in cell-mediated delayed type allergic reactions, but also in antibody-dependent immediate and late type reactions. Proofs for the key role of T lymphocytes in the pathophysiology of allergic drug reactions are positive patch-test reactions and the possibility of the LTT. In the LTT peripheral blood mononuclear cells are obtained from a sensitized patient and cultured in the presence of the suspected drug. Sensitized lymphocytes undergo blastogenesis and generate lymphokines such as IL-2, followed by a proliferative response that can be measured by means of the incorporation of ^{3}H-thymidine during DNA synthesis. The result can be expressed as stimulation index (SI) which is the relation between the cell proliferation with antigen compared without antigen. Recent studies showed that patients with cutaneous hypersensitivity reactions to sulfamethoxazole have sulfonamide-specific peripheral T lymphocytes at a frequency of 1:172 000 [46]. A retrospective evaluation of the sensitivity and specificity of the LTT with a high amount of LTT reactions to β-lactam-antibiotics revealed a sensitivity of 78% and – compared to the patch-test – the same specificity of 85%, whereas the sensitivity of the patch test (64%) was lower than the sensitivity (78%) of the LTT [47].

It has been shown that in penicillin-allergic patients, primarily T helper lymphocytes are induced under the conditions of these test [48]. Even antidrug antibodies, e.g. IgE-antibodies, are under control of T lymphocytes. In order to study the cytokine formation and release, we cloned T lymphocytes from the peripheral blood of patients who suffered from drug allergy, and as an example, Figure 1 shows a CD4$^+$ clone which produced IL-5 and IL-13 but no IFNγ that means a T cell 2 type which is considered to be necessary for a switch to the formation of IgE-antibodies. Such a remarkable IL-5-production has been shown especially in LTT with small molecular weight drugs as antigens [49]. Recently, highly Th2-skewed cytokine profile including IL-4, IL-5 and IL-13, but poor or no expression of IFNγ of β-lactam-antibiotic specific T cells were observed in nonatopic subjects although in other investigations different patterns depending on the molecular structure of the β-lactam-antibiotic derivative, the concentration of the drug and the sign and symptoms of the clinical reaction were described [50–52].

In vitro assays with T lymphocytes may not only be helpful for diagnostic purposes but also to improve our knowledge about the pathophysiology of these reactions. The best evidence for the decisive role of T lymphocytes has come from studies of bullous drug eruptions. Immunohistochemical staining of the skin samples taken from patients suffering from TEN revealed that T lymphocytes which invade the epidermis are more or less exclusively CD8$^+$ cells. We were interested in whether these T lymphocytes are antigen-specific and which function they have [28]. Lesional, drug antigen specific T lymphocytes were cloned from these patients. In order to study the function of these CD8$^+$ cells a cytotoxicity assay was peformed in which the T lymphocytes are incubated with Cr^{51}+-B lymphocytes and a bifunctional antibody which recognized T lymphocytes as well as B lymphocytes. In the case of a cytotoxic activity the B lymphocytes will be destroyed and the radioactivity will be released. This cytotoxic reaction was also shown by incubation of those lymphocytes with keratinocytes [29]. In further studies evidences were provided that this reaction is apoptotic [53] and recently it has been reported that patients suffering from TEN have an increased level of FAS-ligands in the serum, and it was shown that this process is a FAS-FAS-L dependent apoptotic cell death which can be blocked by FAS-L-antibodies as well as by intravenously given immunoglobulines

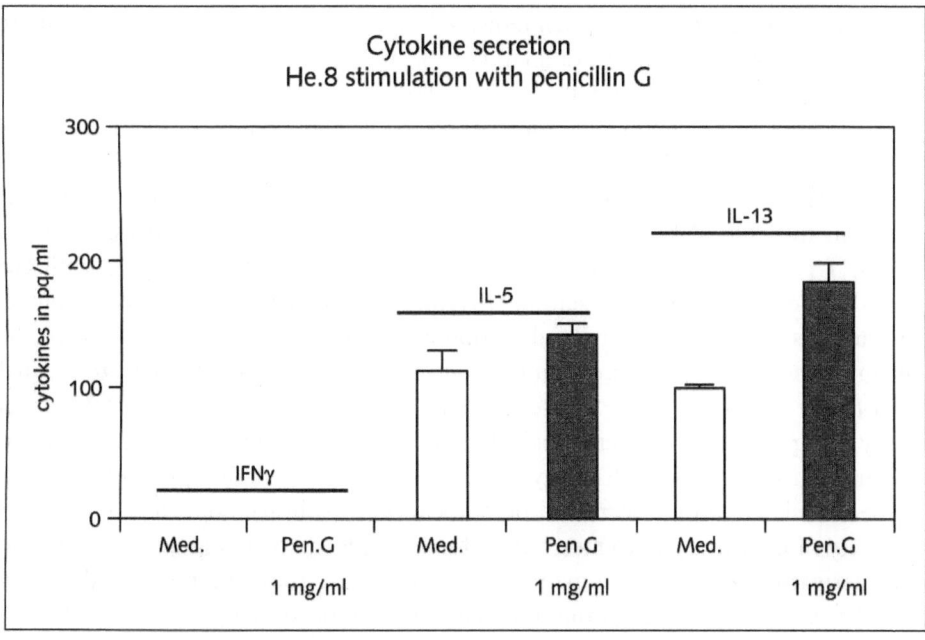

Figure 1
T cell line was obtained from a patient suffering from an morbilliform drug eruption after using ampicillin. IL-5, IL-13, but no IFNγ was released by the lymphocytes [65].

[30]. However this promising observation with regard to new therapeutic options has not yet been confirmed by others.

Although T lymphocytes play a crucial role in drug allergic reactions because the recognition of the drug as an antigen by these cells is a prerequisite for the disease, the role of the LTT especially for drugs which are extensively metabolized is limited. One reason might be that the nominative antigen is not the parent compound but a metabolite of the drug which in most cases is a small molecular weight compound. Those haptens must be bound to high molecular weight compounds in order to be recognized as antigen. However, in order to be bound to proteins those compounds must be chemically activated which can occur by different metabolic pathways such as oxidation by myeloperoxidases or by cytochrome P450-isoenzymes which are present in the liver but also in extrahepatic organs including monocytes, macrophages and dendritic cells [54, 55].

In order to study the influence of such CYP dependent metabolism we modified the already mentioned lymphocyte transformation test by incubating not only with the drug, e.g. metamizole, but also with drug modified microsomal proteins known to contain those CYP's, and found an increased polyclonal proliferation using these drug-modified microsomes, which indicated that such a metabolism may be important for the formation of the nominative antigen [56, 57]. We also cloned lesional T lymphocytes from a patient suffering from bullous drug reaction to sulphonamides and found an increased proliferation measured as cpm of incorporated ^{3}H-thymidine-after adding sulphonamide-modified microsomal protein [58]. Similar results were obtained with anticonvulsants [59–61] Although the influence of drug metabolism in the case of lidocaine or sulfonamides has been questioned by studies in which T cell proliferation was observed after incubation of those drugs with glutaraldehyde fixed antigen presenting cells [62], the importance of drug metabolizing enzymes as a risk factor for the development of drug allergy has been shown for allergic reactions to anticonvulsants by using the lymphocyte toxicity test which lead to the definition of the anticonvulsant hypersensitivity syndrome [59] as well as by using molecular epidemiological methods which showed an increased risk to suffer from sulfonamide-allergy by expressing a slow acetylation phenotype known to be associated with a decreased detoxification of sulfonamides [63]. Finally, the capacity of keratinocytes to metabolize, for example, carbamazepine to highly reactive derivatives in a cytochrome P450 dependent metabolic pathway, and to bind this derivative to proteins of the skin, has been demonstrated [64].

Perspectives

Hypersensitivity reactions are a major problem of drug safety, because they are sometimes influenced by unpredictable individual factors. Further research and better understanding of these individual factors including the study of polymorphically expressed drug metabolizing enzymes will improve drug safety especially if we are able to predict an increased risk of hypersensitivity reaction by appropriate *in vitro* systems. At the level of preclinical drug toxicity testing, a major progress may be the prediction of these reactions, for example, by the application of specifically reacting T lymphocyte clones. The possibility of predicting patients with an increased risk for clozapine-induced agranulocytosis by an *in vitro* cytotoxicity assay by incubation of the parent compound clozapine and the patient's lymphocytes with rat liver microsomes or horseradish peroxidase-peroxide [64] was demonstrated quite recently.

References

1   Davies DM (1985) *Textbook of adverse drug reactions.* Oxford University Press, Oxford
2   Hintner H, Breatnach SM (1993) *Unerwünschte Arzneimittelwirkungen an der Haut.* Blackwell Wissenschaft, Berlin

3   Jäger L, Merk H (1996) *Arzneimittel-Allergie*, Gustav Fischer Verlag, Jena

4   deSwarte RD, Patterson R (1997) Drug allergy. In: R Patterson, LC Grammer, PA Greenberger (eds): *Allergic diseases*, Lippincott Raven, Philadelphia, 317–412

5   Merk HF (1991) Allergische Arzneimittelreaktionen der Haut. *Dtsch med Wschr* 116: 1103

6   Kauppinen K, Alanko K, Hannukelsa M, Maibach H (1998) *Skin reactions to drugs*. CRC Press, Boca Raton

7   Merk HF, Gleichmann E, Gleichmann H (1992) Adverse immunological effects of drugs and other chemicals and methods to detect them. In: K Miller, J Turk, S Nicklin (eds): *Principles and practice of immunotoxicology*, Blackwell, 86–103

8   Merk H, Jugert F, Frankenberg S (1993) Keratinocytes and reconstructed epidermis for *in vitro* pharmacological and toxicological testing. In: V Rogiers, W Sonck, E Shephard, A Vercruysse (eds): *Human cells in in vitro pharmaco-toxicology*. VUB Press, Brüssel, 45–59

9   Wolkenstein P, Tan C, Lecoeur S, Wechsler J, Garcia-Martin N, Charue D, Bagot M, Beaune Ph (1998) Covalent binding of carbamazepine reactive metabolites to P450 isoforms present in the skin. *Chem Biol Interactions* 113: 39–50

10   Weiss ME (1992) Drug allergy. *Med Clin North Am* 76: 857–882

11   Weiss ME, Adkinson NF (1988) Immediate hypersensitivity reactions to penicillin and related antibiotics. *Clin Allergy* 18: 515–540

12   Mockenhaupt M, Schöpf E (1998) Epidemiology of severe cutaneous drug reactions. In: K Kauppinen, K Alanko, M Hannukelsa, H Maibach (eds): *Skin reactions to drugs*. CRC Press, Boca Raton, 3–15

13   Atkinson TP, Kaliner MA (1992) Anaphylaxis. *Clin Allergy* 76: 841–855

14   Merk H, Eichler G (1995) Anaphylaktoide Reaktionen. In: G Plewig, H Korting (eds): Fortschritte der Dermatologie, Band 14, Springer-Verlag, Heidelberg, 108–115

15   Weiss ME, Nyan D, Zhiang P, Adkinson NF (1989) Association of protamine IgE and IgG antibodies with life-threatening reactions to intravenous protamine. *N Engl J Med* 320: 886

16   Merk H, Goerz G (1983) Analgetika-Intoleranz. *Z Hautkr* 58: 535–542

17   Dahlen B, Dahlen SE (1995) Intolerance reactions to NSAIDs. In: A Basomba, J Sastre (eds): *Proceedings of XVI. European congress of allergology and clinical immunology*, Monduzzi Editore, Bologna, 821–828

18   Drazen JM, Israel E, O'Byrne PM (1999) Treatment of asthma with drugs modifying the leukotriene pathway. *N Engl J Med* 340: 197–206

19   Huwyler T, Wüthrich B, Mühletaler K, Kuhn H, Jungbluth H, Späth P, Hochreutener H (1989) Enalapril assoziierte Angioödeme. *Schweiz Med Wschr* 119: 1253

20   Charlesworth EN, Hood AF, Soter NA, Kagey-Sobotka A, Norman PS, Lichtenstein LM (1989) Cutaneous late-phase response to allergen. *J Clin Invest* 83: 1519

21   Merk H, Ruzicka T (1981) Oral contraceptiva as a cause of erythema nodosum. *Arch Dermatol* 117: 454

22   Hannuksela M (1998) Systemic contact dermatitis. In: K Kauppinen, K Alanko, M Hannukelsa, H Maibach (eds): *Skin reactions to drugs*. CRC Press, Boca Raton, 51–58

23   Huff JC, Weston WL, Tonnesen MG (1983) Erythema multiforme: A critical review of characteristics, diagnostic criteria, and causes. *J Am Acad Dermatol* 8: 763–775

24   Kauuppinen K, Kariniemi AL (1998) Clinical manifestations and histological characteristics. In: K Kauppinen, K Alanko, M Hannukelsa, H Maibach (eds): *Skin reactions to drugs*. CRC Press, Boca Raton, 25–50

25   Shiohara T, Nickoloff BJ, Sagawa Y (1998) Fixed drug eruption. Expression of epidermal keratinocyte intercellular adhesion molecule-1 (ICAM-1). *Arch Dermatol* 125: 1371

26   Roujeau JG, Stern RS (1994) Severe adverse cutaneous reactions to drugs. N Engl J Med 331: 1272–1285

27   Ruojeau JC, Kelly JP, Naldi L, Rzany B, Stern RS, Anderson T, Auquier A, Bastuji-Garin S, Correia O, Locati F et al (1995) Medication use and the risk of Stevens-Johnson Syndrome or toxic epidermal necrolysis. *N Engl J Med* 333: 1600–1607

28   Hertl M, Merk HF (1995) Lymphocyte activation in cutaneous drug reactions. *J Invest Dermatol* 105 (Suppl. 1): 95S–98S

29   Hertl M, Bohlen H, Jugert F, Boecker C, Knaup R, Merk HF (1993) Predominance of epidermal CD8⁺ T lymphocytes in bullous reactions caused by β-lactam antibiotics. *J Invest Dermatol* 101: 794–799

30   Viard I, Wehrli Ph, Bullani R, Schneider P, Holler N, Salomon D, Hunziker Th, Saurat J-H, Tschopp J, French LE (1998) Inhibition of toxic epidermal necrolysis by blockade of CD95 with human intravenous immunoglobulin. *Science* 282: 490–493

31   Ippen H (1993) Photoxische und photoallergische Reaktionen: Möglichkeiten der Risikoabschätzung in der präklinischen und klinischen Entwicklung. In: HF Merk (Hrsg): *Allergische und pseudoallergische Arzneimittelreaktionen: Stellenwert von Epidemiologie und zellulären Testsystemen in Diagnostik und Vorbeugung*. Blackwell Wissenschaft, Berlin, 37–45

32   Elmets CA: Cutaneous phototoxicity. In: HW Lim, NA Soter (eds) (1993) Clinical photomedicine, Marcel Dekker, New York, 207–226

33   deLeo VA (1993) Photoallergy. In: HW Lim, NA Soter (eds.): *Clinical photomedicine*, Marcel Dekker, New York, 227–239

34   Spielmann H, Lovell WW, Hölzle E, Johnson BE, Maurer Th, Miranda MA, Pape WJW, Sapora O, Sladowski D (1994) *In vitro* photoxicity testing. *ATLA* 22: 314–348

35   Merk HF, Gleichmann E, Gleichmann H (1992) Adverse immunological effects of drugs and other chemicals and methods to detect them. In: K Miller, J Turk, S Nicklin (eds): *Principles and practice of immunotoxicology*, Blackwell, Oxford, 86–103

36   Totoritis MC, Tan EM, McCally EM, Rubin RL (1988) Association of antibody to histone complex H2A-H2B with symptomatic procainamide-induced lups. *N Engl J Med* 318: 1431–1436

37   Hess E (1988) Drug induced lupus. *N Engl J Med* 318: 1460–1462

38   Kretz-Rommel A, Rubin RL (1999) Persistence of autoreactive T cell drive is required

to elicit anti-chromatin antibodies in a murine model of drug-induced lups. *J Immunol* 162: 813–820

39   Chandler MJ, Grammer L, Patterson R (1987) Provocative challenge with local anesthetics in patients with a prior history of reactions. *J Allergy Clin Immunol* 79: 883–886

40   Cuesta-Herranz J, Heras de las M, Fernandes M, Lluch M, Figueredo E, Umpierrez A, Lahoz C (1997) Allergic reaction caused by local anesthetic agents belonging to the amide group. *J Allergy Clin Immunol* 99: 427–428

41   Gall H, Kaufmann R, Kalveram CM (1996): Adverse reactions to local anesthetics; analysis of 197 cases. *J Allergy Clin Immunol* 97: 933–937

42   Matsson P, Ahlstedt S, Venge P, Thorell J (1991) *Clinical impact of the monitoring of allergic inflammation*. Academic Press, London

43   Furukawa K, Tengler R, de Weck AL, Maly FE (1994) Simplified sulfidoleukotriene ELISA using LTD4-conjugated phosphatase for the study of allergen-induced leukotriene generation by isolated mononuclear cells and diluted whole blood. *J Invest Allergol Clin Immunol* 4: 110–115

44   Berg PA, Becker EW (1993) Zelluläre Immunreaktionen bei Patienten mit medikamentösen Nebenwirkungen: Diagnostische Relevanz des Lymphozytentransformationstestes. In: H Merk (Hrsg): *Allergische und pseudoallergische Arzneimittelreaktionen*, Blackwell Wissenschaft, Berlin, 102–118

45   Pichler WJ (1993) Der Lymphozytentransformationstest in der Diagnostik von unerwünschten Arzneimittelreaktionen. In: HF Merk (Hrsg): *Allergische und pseudoallergische Arzneimittelreaktionen: Stellenwert von Epidemiologie und zellulären Testsystemen in Diagnostik und Vorbeugung*, Blackwell Wissenschaft, Berlin, 119–130

46   Kalish RS, LaPorte A, Wood JA, Johnson KL (1994) Sulfonamide-reactive lymphocytes detected at very low frequency in the perpheral blood of patients with drug-induced eruptions. *J Allergy Clin Immunol* 94: 465–472

47   Nyfeler B, Pichler WJ (1997) The lymphocyte transformation test for the diagnosis of drug allergy: sensitivity and specificity. *Clin Exp Allergy* 27: 175–181

48   Koponen M, Pichler WJ, De Weck AL (1986) T cell reactivity to penicillin: Phenotypic analysis of *in vitro*-activated cell subsets. *J Allergy Clin Immunol* 78: 645

49   Pichler WJ, Zanni M, Greyers S, Schnyder B, Mauri-Hellwig D, Wendland T (1997) High IL-5 production by human drug-specific T cell clones. *Int Arch Allergy Immunol* 113: 177–180

50   Brugnolo F, Annunziato F, Sampognaro S, Campi P, Manfredi M, Matucci A, Blanca M, Romagnani S, Maggi E, Parronchi P (1999) Highly Th2-skewed cytokine profile of β-lactam-specific T cells from nonatopic subjects with adverse drug reactions. *J Immunol* 163: 1053–1059

51   Merk HF, Hertl M (1996) Immunologic mechanisms of cutaneous drug reactions. *Sem Cut Med Surg* 15: 228–235

52   Padovan E, Greyerz S, Pichler WJ, Weltzien HU (1999) Antigen-dependent and independent IFNγ modulation by penicillins. *J Immunol* 162: 1171–1177

53   Paul C, Wolkenstein P, Adle H, Wechsler J, Garchon HJ, Revuz J, Roujeau JC (1996)

Apoptosis as a mechanism of keratinocyte death in toxic epidermal necrolysis. *Brit J Dermatol* 134: 710–714

54   Baron JM, Zwadlo-Klarwasser G, Jugert F, Hamann W, Rübben A, Mukhtar H, Merk HF (1998) Cytochrome P450 1b1 is the major P450 isoenzyme in human blood monocytes and macrophage subsets. *Biochem Pharmacol* 56: 1105–1110

55   Sieben S, Baron JM, Blömeke B, Merk HF (1999) Multiple cytochrome P450-isoenzymes mRNA are expressed in dendritic cells. *Int Arch Allergy Immunol* 118: 358–361

56   Merk H, Schneider R, Scholl P (1987) Lymphocyte stimulation by drug-modified microsomes. In: RW Estabrook, E Lindenlaub, F Oesch, AL de Weck (eds): *Toxicological and immunological aspects of drug metabolism and environmental chemicals.* FKSchattauer Verlag, Stuttgart, New York

57   Merk HF, Baron J, Hertl M, Niederau D, Rübben A (1997) Lymphocyte activation in allergic reactions elicited by small-molecular-weight compounds. *Int Arch Allergy Immunol* 113: 173–176

58   Hertl M, Jugert F, Merk HF (1995) CD8+ dermal T cells from a sulphamethoxazole-induced bullous exanthem proliferate in response to drug-modified liver microsomes. *Brit J Dermatol* 132: 215–220

59   Shear NH, Spielberg SP (1988) Anticonvulsant hypersensitivity syndrome. *In vitro* assessment of risk. *J Clin Invest* 82: 1826–1832

60   Friedman PS, Pirmohamed M, Park K (1994) Investigations of mechanisms in toxic epidermal necrolysis induced by carbamazepine. *Arch Dermatol* 130: 598–604

61   Gall H, Merk H, Scherb W, Sterry W (1994) Anticonvulsiva-Hypersensitivitäts-Syndrom auf Carbamazepin. *Hautarzt* 45: 494–498

62   Pichler WJ, Schnyder B, Zanni MP, von Greyerz S (1998) Role of T cells in drug allergies. *Allergy* 53: 225–232

63   Wolkenstein P, Carriere V, Charue D, Bastuji-Garin S, Revuz J, Roujeau JC, Beaune P, Bagot M (1995) A slow acetylator genotype is a risk factor for sulhonamide-induced toxic epidermal necrolysis and Stevens-Johnson syndrome. *Pharmacogenetics* 5: 255–258

64   Tschen AC, Rieder MJ, Oyewumi LK, Freeman DJ (1999) The cytoxicity of clozapine metabolites: implications for predicting clozapine-induced agranulocytosis. *Clin Pharmacol Ther* 65: 526–532

65   Sachs B, Merk HF, Hertl M (1999) T cell reactivity to β-lactam antibiotics. *Arch Dermatol Res* 291: 169A

Future developments in treatment of atopic dermatitis

Birgit A. Pees and Peter S. Friedmann

Dermatopharmacology Unit, Level F, South Block, Southampton General Hospital, Southampton, SO16 6YD, UK

Introduction

In recent years there have been major advances in both the understanding of basic mechanisms underlying atopic dermatitis and in the development of drugs with specific effects on different components of immune or inflammatory processes. These drugs form agents which can be used on the one hand to treat atopic dermatitis, and on the other, they represent tools which can help dissect the immuno-pharmacological mechanisms involved in the process. These agents will be studied over the next few years and some will come to form a new approach to the treatment of atopic dermatitis.

Immunosuppressives

This group of drugs includes Cyclosporin A (CyA), FK506, ascomycin and rapamycin. These drugs came to prominence through their use in the field of transplantation immuno-suppression. Following the chance observations that various skin diseases such as psoriasis improved in transplant recipients treated with these agents, it was not long before they were used intentionally in the treatment of a variety of inflammatory diseases. CyA has made an immense impact on the treatment of atopic dermatitis in the last few years. However, the problems of significant side-effects accompanying the need for it to be given systemically, mean that its use is always going to be limited. FK506 (Tacrolimus) is set to take on a much greater role by virtue of the fact that it is effective when used topically. Work is in progress to bring both ascomycin and Rapamycin derivatives to the market for use in inflammatory skin diseases. It is now known that these drugs work through very similar mechanisms which suppress transcription of a range of molecules such as cytokines, which are critical in the normal activation of T cells. Hence they will be considered together although CyA and FK506 will be the main focus.

Immunology and Drug Therapy of Allergic Skin Diseases, edited by C.A.F.M. Bruijnzeel-Koomen and E.F. Knol
© 2000 Birkhäuser Verlag Basel/Switzerland

Production of cytokines during the immune response is inhibited by Cyclosporin A (CyA) and FK506. T lymphocytes were the first identified targets of these drugs which were shown to inhibit cytokine gene transcription. Since then it has become clear that other cell types including B cells and mast cells are also inhibited by CyA. These observations suggested the existence of a common, CyA-sensitive pathway for regulation of inducible cytokine genes in a range of cell types.

The molecular target for both drugs is calcineurin, a calcium/calmodulin dependent serine/threonine phosphatase [1]. Calcineurin appears to have as its main target a transcription factor called NFAT (nuclear factor of activated T cells) which occurs in several forms: NFAT1 includes types 1a previously called NFATp, and type 1c previously called NFATc [2]. NFAT1 is found mainly in lymphoid cells (T and B lymphocytes, NK cells, monocytes, some macrophages and in murine bone marrow-derived mast cells) and human umbilical vein endothelial cells [3]. NFAT proteins are activated by stimulation of receptors coupled to calcium mobilisation – such as antigen receptors on T and B cells, Fcε receptors on mast cells and basophils, Fcγ receptors on macrophages and NK cells. Following receptor stimulation and calcium mobilisation, many intracellular enzymes are activated including calcineurin. Calcineurin removes phosphate groups from NFAT, exposing the nuclear localisation signal which causes the NFAT to rapidly translocate from the cytoplasm to the nucleus. The NFAT targets the promoter regions of many genes including transcription factors, signalling proteins, cytokines, surface receptors and others which are inducibly transcribed following cellular activation.

CyA and FK506 enter cells readily and bind to different classes of intracellular receptors called immunophilins. The receptors for CyA are cyclophilins and for FK506 the 12Kda FK-binding protein (FKBP-12). Although these are structurally different proteins, they both have activity as peptidyl prolyl *cis-trans* isomerase otherwise known as rotamase [1]. When the drugs – CyA or FK506 – are bound to their respective immunophilin, the rotamase activity is inhibited. It was originally thought that the immuno-suppressive actions of these drugs were a consequence of inhibition of rotamase. However, it is now clear that this is not the case. Instead, the complex of immunophilin plus drug inhibits the phosphatase activity of calcineurin and thereby, blocks the activation and nuclear translocation of NFAT and subsequent transcription of its target genes (Fig. 1). There is a direct proportionality between the *in vitro* phosphatase inhibiting properties of these drugs and their *in vivo* immuno-suppressive potency reflected by suppression of cytokine production.

CyA has been shown in many clinical trials to be of great therapeutic efficacy in the treatment of atopic dermatitis [4–8]. However, as mentioned above, its usefulness is limited by the need for oral administration which is accompanied by a wide range of side-effects. There is thus a great need for alternative agents that are effective when used topically.

Tacrolimus inhibits transcription of IL-2 and many other cytokines by T and B lymphocytes; it inhibits transcription of the IL-8 receptor gene in keratinocytes. It

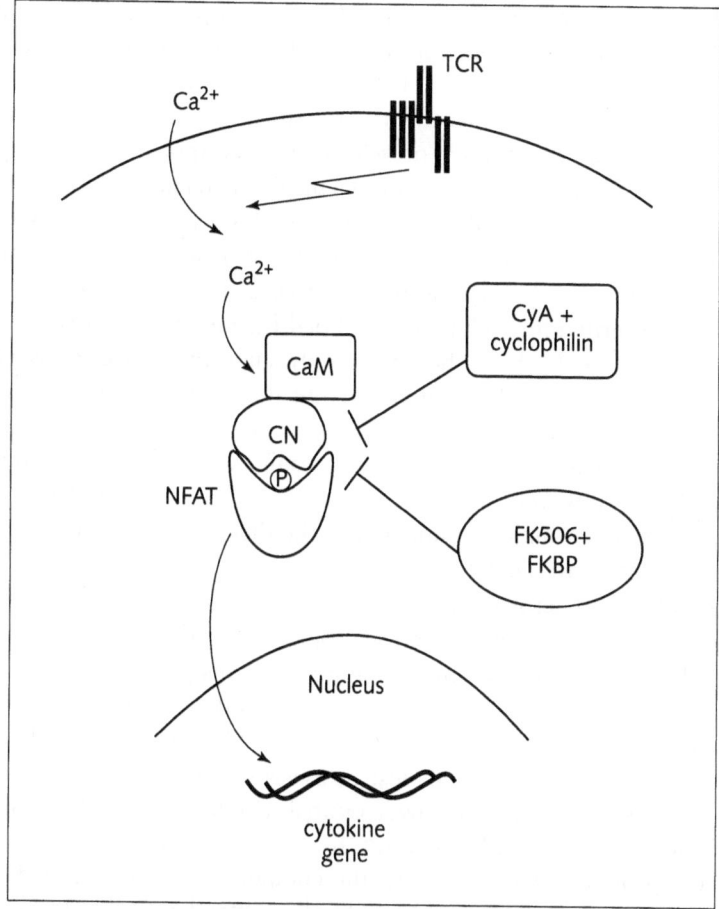

Figure 1
NFAT proteins are activated by stimulation of receptors coupled to calcium mobilisation -
such as the T cell receptor (TCR). Calcium/calmodulin (CaM) activates calcineurin (Cn). Cal-
cineurin removes phosphate groups from NFAT, exposing the nuclear localisation signal
which causes the NFAT to rapidly translocate from the cytoplasm to the nucleus. The NFAT
targets the promoter regions of many genes. The immunosuppressive drugs CyA or FK506,
complexed with specific immunophilins, inhibit the activation of calcineurin, and hence, the
nuclear translocation of NFAT.

reduces anti-IgE-induced histamine release by mast cells *in vitro*. It inhibits or down-
regulates expression of Fcε R1 and CD80 (B7.1) but not CD86 (B7.2) by epidermal
Langerhans cells [9]. This is accompanied by inhibition of some antigen presenting
functions.

A number of clinical trials have examined the efficacy of topical Tacrolimus in atopic dermatitis. Some of these were open studies [10–12], while others have been double blind placebo controlled studies [13, 14]. In the largest multi-centre study, 200 adult patients were treated for 3 weeks with one of three different concentrations - 0.03, 0.1 and 0.3% of Tacrolimus, or vehicle control, applied twice daily. Highly significant clinical benefits were seen with the median percentage decrease in scores being up to 83% [14]. In a multi-centre, double-blind, placebo-controlled trial the effects of the same concentrations of Tacrolimus ointment were also examined in treatment of childhood atopic dermatitis for 3 weeks. There was up to 70% improvement in global severity score while the modified Eczema Area and Severity Score improved by up to 81% [13]. As there were only slight differences between the different concentrations it is probable that it is not necessary to use strengths greater than 0.1%. At present other multi-centre trials of Tacrolimus are under way in longer term treatment of atopic dermatitis and results will become available towards the end of 2000. However, the initial results look so impressive that it is clear Tacrolimus will become a major component in the therapeutic armamentarium available for treatment of atopic dermatitis.

Derivatives of Ascomycin macrolactams also have immunomodulatory and anti-inflammatory effects. They are effective when used topically against allergic contact hypersensitivity (ACH) reactions in a number of animal systems [15]. For example, in mice or pigs, 0.1% Ascomycin (SDZ ASM 981) was equivalent to clobetasol-17-propionate, and inhibited ACH responses by up to 60% [15]. This compound has been shown to have therapeutic efficacy against psoriasis in a micro plaque assay [16]. Small areas of psoriasis in 15 patients were treated with 0.1 or 1% SDZ ASM 981 or a potent steroid or placebo with occlusion. By 11 days the Ascomycin derivative had cleared the psoriasis. More recently, the compound has been tested in treatment of atopic dermatitis in a double-blind placebo controlled trial [17]. Symmetrical local areas on either side of the body were treated once or twice daily for 21 days. The eczema was scored at regular intervals. Twice daily applications resulted in a 70% reduction in the eczema score by day 16, compared with 43% reduction after once daily application and 15% improvement with placebo. There was some clinical effect after only 2 days. There were no side-effects from the cream application. Clearly this compound requires further evaluation and trials are in progress to examine its efficacy and safety profile.

Rapamycin (sirolimus), like Tacrolimus, binds to the FKBP-12. However, the complex apparently does not inhibit transcription of cytokines. Instead, it inhibits activation of cellular proliferation resulting from interaction of IL-2 with its receptor (IL-2R). When applied topically, Rapamycin had moderate efficacy against irritant contact dermatitis [18]. Interestingly, when used either topically or systemically in experimental contact dermatitis, whereas Tacrolimus was highly potent, Rapamycin was without effect [18]. The role of Rapamycin has not yet been assessed with regard to use in inflammatory dermatoses.

Mycophenolate mofetil

This agent is finding a place in the field of transplantation immunosuppression. It is an ester prodrug which is rapidly converted to mycophenolic acid (MPA). MPA is a potent, reversible, non-competitive inhibitor of inosine monophosphate dehydrogenase. This is an important enzyme in the conversion of inosine monophosphate to guanosine monophosphate, required for purine biosynthesis. MPA causes a reduction in GTP and other guanine nucleotides in lymphocytes but not neutrophils. This results in inhibition of DNA synthesis and other GTP-dependent metabolic events [19] in T and B lymphocytes but not other non-lymphoid cells. The safety profile is generally good. The major side-effects of MPA are gastro-intestinal upsets. However, nephrotoxicity and hepatotoxicity do not seem to occur although there may be some bone marrow suppression. This agent is now being explored in a range of auto-immune and inflammatory diseases including rheumatoid arthritis [20]. There are reports of its use in bullous pemphigoid [21] and psoriasis [22]. It is an obvious candidate for future assessment in atopic dermatitis.

In conclusion, the immuno-suppressant drugs from this group that appear to be active when administered topically, will be undergoing extensive evaluation in the near future. It looks highly probable that some at least – Tacrolimus and probably ascomycin derivatives – will become a major part of the future therapies used for atopic dermatitis.

Phosphodiesterase inhibitors

For several years an area of the basic biology of the atopic state that has been developing is that of altered signalling *via* cyclic AMP-dependent pathways. The disturbances, summarised below, indicate that in atopic individuals attenuated cAMP signalling results from overactive cAMP-phosphodiesterase which is associated with several of the phenotypic features. Following the development of reasonably potent and specific inhibitors of phosphodiesterase (PDE) there has been a resurgence of interest in their possible use in the treatment of atopic diseases including eczema and asthma.

The principal observations summarised by Hanifin [23] follow from the well known increased IgE synthesis that characterises atopic individuals. *In vitro* this appears linked to reduced IFNγ production by mixed PBMC from patients with atopic dermatitis. However, atopic T cells actually show increased production of IFNγ, hence it was concluded the monocytes produce an inhibitor of IFNγ synthesis. Atopic PBMC produce increased amounts of prostaglandin E_2 (PGE_2) and it is the monocytes which are the source [24]. It had been shown by Betz and Fox that PGE_2 inhibits IFNγ production [25] and Chan et al. showed that inhibition of PGE_2 with Indomethacin results in increased IFNγ production by atopic PBMC [24].

Atopic monocytes also produce increased amounts of IL-10 which also inhibits production of IFNγ [17].

Hanifin had previously shown that PBMC from atopic subjects have blunted cAMP responses to agents acting on adenylyl cyclase such as histamine, iso-prenaline or prostaglandins [26]. This was found to be a result of elevated cAMP-phosphodiesterase (PDE) activity [27], so the cAMP was hydrolysed more completely. These abnormalities were present in people with respiratory atopic disease too. Sawai et al. [28] showed the raised PDE activity was present in children, showed no correlation with disease severity and was elevated when patients went into remission. The abnormality has also been shown to be present in cord blood from babies with atopic parents [29, 30].

The elevated cAMP-PDE activity was located predominantly in monocytes [31]. Chan and colleagues went on to type the PDEs in atopic leukocytes [32]. They found three isoforms that were common to both monocytes and lymphocytes and all were more active than in cells from non-topics. However, the atopic monocytes had an additional isoform which was specific for cAMP and was calcium/calmodulin dependent. Also, the PDE was more susceptible than control PDE to inhibition by Ro20-1724 – an inhibitor of PDE type 4 [32, 33]. It should be noted that Gantner et al. found no differences between PDE expression or activity in cells from healthy or atopic blood donors [34].

One aspect which makes interpretation of data about levels of PDE activity difficult is that PDE activity is subject to modulation by various factors. Thus, PDE activity is induced by stimuli which increase cAMP, such as long-term β adrenergic agonists. This has clouded the question of whether elevated PDE is found in atopic asthma since most patients receive chronic therapy with these agents. In monocytes from healthy subjects PDE activity is increased by cytokines IL-4 and IFNγ [35]. However, in monocytes from AD patients, IFNγ at low concentrations was without effect and at higher concentrations it reduced PDE activity [35]. This difference may be because in atopic cells PDE activity is already maximally induced.

The PDEs form a family of at least seven groups (Tab. 1) [36]. The different types vary in their affinity for cAMP or cGMP, their susceptibility to different inhibitors and in some cases, in their cell/tissue distribution. Type 4 PDE (PDE4) is cAMP-specific and was characterised by inhibition by rolipram and Ro20-1724. PDE type 4 is present in inflammatory cells: mast cells, eosinophils, monocytes, macrophages and lymphocytes.

Several activities of mononuclear cells are mediated *via* cAMP and can be reduced by inhibitors of PDE4: increased PGE_2 production [17]; spontaneous IgE production by B cells [37] and release of histamine from basophils [17]; increased spontaneous production of IL-10 is reduced as is the anti-CD3 induced increase in IL-4 production [17, 38]; migration of eosinophils [39]; *in vitro* trans-migration of lymphocytes but not monocytes through a monolayer of endothelial cells [40]; allergen-induced release of IL-5 [41].

Table 1 - Classification of phosphodiesterase isoenzymes

Family	Characteristic	Tissue	Selective inhibitors
1A-D	Ca/Calmodulin stimulated	Brain, lung	Vinpocetine
2	cGMP-stimulated	heart, vascular smooth muscle, airway smooth muscle; platelets	EHNA
3A,B	cGMP-inhibited	Ditto	siguazodan milrinone cilostamide
4A-D	cAMP-specific	Inflammatory cells, airway smooth muscle, heart, brain	rolipram, Ro20-1724, denbufylline, CP80633
5	cGMP-specific	Trachea, aorta, platelets	zaprinast
6	photoreceptor	Retinal rods and cones	zaprinast
7	cAMP-specific (high affinity)	lymphocyte	??

(Modified from Spina et al. [36])

Phosphodiesterase inhibitors

Theophylline is probably the best known of the PDE inhibitors. For many years it has been used in the treatment of bronchial asthma. The supposed rationale behind its use was that, by inhibition of PDE, cAMP responses are boosted when receptors such as β-adrenoceptors are stimulated. Moreover, administration of theophylline mimics adrenergic stimulation, resulting in bronchodilatation. However, theophylline inhibits the late asthmatic response and has a number of effects which are "anti-inflammatory". Thus it reduces IL-2 release from lymphocytes, oxygen radical generation and cytokine release from monocyte/macrophages and arachidonic acid release from eosinophils [42]. Low-dose administration of theophylline for up to 6 weeks in asthma resulted in significant diminution of a range of aspects of the cellular inflammatory response [36].

Type 4 PDE inhibitors

A number of compounds are under investigation as potential treatments of allergic disorders including asthma and atopic dermatitis (Tab. 1). Their systemic use is significantly hampered by the side-effects of nausea and headache – probably reflecting the presence of PDE4 in the brain. However, the possibility of treating atopic dermatitis by topical administration allows achievement of high local concentrations without significant systemic side-effects. Hanifin et al. [17] reported the results of a double-blind, placebo-controlled trial of topical CP80233, an inhibitor of PDE4. This compound was chosen because of its *in vitro* potency. Twenty patients with atopic dermatitis used 0.5% CP80233 on one side and petrolatum vehicle on the other side. Treatment was applied to 200 cm^2 areas of lesional skin for 28 days. An improvement in clinical score was observed in 16 of 20 sites receiving active treatment compared with 3 of 20 placebo treated sites (p < 0.001). The clinical improvement was apparent after only 3 days and increased further by 28 days. This was interesting because, in contrast to the effect of theophylline in asthma, there was no evidence of tachyphylaxis. Unfortunately, the magnitude of clinical improvement was not very great. Clearly further trials of this and other compounds are needed to assess the potential value of this class of drugs in the treatment of atopic dermatitis. Also, the clinical use of these agents should help to establish the biological significance of the alterations in cAMP signalling pathways in the maintenance of the clinical pathology.

Leukotriene antagonists

The leukotrienes are arachidonic acid metabolites formed by the action of 5-lipoxygenase. The sulphido-peptide leukotrienes LTC$_4$, LTD$_4$ and LTE$_4$ together make up the slow reacting substance of anaphylaxis. They are released from granulocytes – neutrophils, eosinophils and basophils – as well as mast cells. Antigen-mediated cross-linking of surface IgE is one of the principal physiological stimulants for their release and their synthesis is inhibited by corticosteroids. The leukotrienes are vasoactive, causing vasodilatation and increased permeability, as well as having chemotactic and smooth muscle contracting properties. They are of major pathological significance in the "late phase" asthmatic response developing from 4–6 h after exposure to allergen. They are thought to be involved in the late-phase response developing in skin after intradermal inoculation of allergen [43] but their presence and importance in atopic dermatitis is far from clear. LTC$_4$ has been detected in suction blister fluid from atopic dermatitis [44]. Fauler and colleagues suggested that people with atopic dermatitis excrete increased amounts of leukotrienes in the urine [45], but Sansom et al failed to confirm this observation [46]. The advent of drugs which either inhibit synthesis of leukotrienes or block their interaction with specific receptors, will now allow dissection of the role of these mediators in the pathogenesis of atopic dermatitis. The leukotriene receptor antagonist Zafirlukast has been reported to induce significant clinical benefit in patients with atopic dermatitis [47]. However, this was an open, pilot study of only four patients and clearly confirmation with appropriately controlled trials will be required to establish the position of these drugs in the future treatment of atopic dermatitis.

Ultraviolet therapy

Ultraviolet light has long been used in the treatment of atopic dermatitis. Recently, new lamps and methods of treatment have been explored.

In contrast to psoriasis [48], the UV action spectrum for clearing AD has not yet been defined. Hence, there is no general consensus as to which portion of the UV spectrum is most effective to treat this condition. The UV modalities for the treatment of AD which have been studied include

(1) broadband UVB (290–320nm) [49–51];
(2) solar simulated combined UVA and UVB (300–400 nm) [50–52];
(3) narrow band UVB (311 ± 2 nm) [53, 54];
(4) high-dose UVA1 (340–400 nm) [55–57];
(5) photochemotherapy (PUVA) using a chemical photosensitiser, e.g. 8-methoxy-psoralen and UVA (320–400 nm) [58, 59].

PUVA has been used successfully to treat children and adolescents with severe AD [60, 61] with long-term remissions in some of the patients but with clear restrictions for very severe cases because of the long-term carcinogenic risks [62]. The tolerability of UV-therapy varies considerably depending on the patient's sensitivity and the UV dose regimen applied. Thus modifications of incremental regimens for UVB therapies have been examined [49, 51].

In conventional UVB therapy dose-regimens are employed which are often associated with side-effects such as erythema, xerosis and pruritus and some patients even experience exacerbation of their eczema. Hence, Wulf et al. [51] developed a very low-dose increment protocol guided by skin reflectance measurement in order to decrease the cumulative dose and yet maintain efficacy. In a retrospective open study 15 patients who received the low increment regimen were compared with 17 patients treated with the standard regime (12.5–25% every second irradiation). There was no significant difference in duration of treatment or occurrence of erythema but for the low increment regimen the healing score was significantly higher. Further, the cumulative UVB dose was four times lower with a mean total increase of the exposure dose of only 20% during treatment without loss of efficacy for the low-dose group. Jekler et al. [49], in a half-body study of 25 patients, showed no significant difference in clinical efficacy between irradiation regimens using 0.4 times the minimal erythema dose (MED) *versus* 0.8 MED. These findings suggest that low-dose UVB therapy for AD can be performed without losing effect. The reduction of cumulative dose will be important regarding concerns abut long-term safety [63, 64].

Philips TL01 fluorescent lamps emit narrow band UVB (311 ± 2 nm). This has been shown to be effective in the treatment of psoriasis and, compared with broadband UVB, appears to be more effective with fewer side-effects such as erythema

[65]. Compared to broadband UVB these lamps have a decreased emission of more erythemogenic wavelengths between 290 and 305 nm and a five to six-fold increased emission of the longer UVB wavelengths around 312 nm [53].

These lamps have now been shown to have clinical efficacy against severe AD. George et al. performed an open study with 21 patients and showed a significant reduction of severity and use of topical steroids [53]. TL01 UVB is also effective in children with AD [54]. The authors reported a good response in the majority of patients and amelioration of pruritus, restoration of a normal sleeping pattern and reduction of topical steroid use as the main benefits described by parents and patients. However, truncal and facial erythema occurred as an adverse effect in many patients, which could be attributed to the irradiation regimen of 10–20% increase each treatment (three times weekly). Further controlled studies are needed to evaluate different irradiation regimens as for broadband UVB [49, 51].

Another novel phototherapy modality, long-wave UVA1 (340–400 nm), has proved to be therapeutically effective as monotherapy for the treatment of severe acute exacerbations of AD [55, 57]. Based on studies which indicate that a combination of UVA and UVB increases therapeutic efficacy as compared with UVB alone [50, 52], Krutmann et al. investigated the effectiveness of high-dose UVA1, which allows doses of 130 J/cm^2/day [55, 57]. In a randomised multi-centre study, 53 patients received either UVA1, UVA/B or topical fluocortolone therapy daily for a total of 10 days. UVA1 was superior to both topical steroid and the UVA/B. It was also observed that blood eosinophilia and eosinophil cationic protein (ECP) levels (a marker for disease activity in AD) were markedly reduced by UVA1 and topical steroid but remained essentially unaltered in patients treated with UVA/B [57]. Grabbe et al. demonstrated that clinical improvement after UVA1 but not UVA/B therapy was closely linked with immunomodulatory effect in the skin [56]. UVA1 appears to reduce the numbers of dermal Langerhans cells and mast cells as well as the relative numbers of epidermal IgE$^+$ Langerhans cells despite increased numbers of CD1a$^+$ cells.

Although UVA1 is effective, the high-dose regimen raises questions regarding its long-term safety. Therefore it is recommended as short courses for patients with acute exacerbations of AD [57].

Finally, extracorporeal UVA photopheresis [66] has been used successfully as rescue therapy in three patients with severe AD who had intractable disease and failed to respond to conventional first line and immunosuppressive therapy [67]. Photopheresis was performed every 2 weeks on 2 consecutive days and this cycle was repeated 10 times. All three patients responded well to treatment and two patients stayed in remission for 8 and 12 months. A reduction of elevated total IgE and ECP levels could also be demonstrated. These findings suggest that photopheresis could be a useful tool in the treatment of severe AD but it requires highly specialised equipment and is unlikely to become a widely available treatment.

Traditional Chinese herbal therapy

In recent years there has been considerable interest in traditional Chinese herbal therapy (TCHT). The problem with assessing its usefulness is that the traditional Chinese approach is polypharmaceutical and treatment is tailored to each individual.

The 10-herb mixture of Dr Luo, a Chinese practitioner collaborating in the controlled trials, has become widely known to western dermatologists and several studies have been conducted to analyse the ingredients of this mixture and to investigate the pharmacological properties of the herbal ingredients [68]. It has been found that all herbs had non steroidal anti-inflammatory effects and in addition some had antihistaminic or steroid-like properties; one herb appeared to act as an immunosuppressant [69]. Clinical efficacy in atopic eczema patients treated with Zemaphtye (commercial 10-herb mixture) seems to be associated with immunological changes such as decrease of serum IgE, vascular cellular adhesions molecules and IL-2R [70].

Recently, the effect of TCHT has been studied *in vitro* [71]. It was found that an extract of traditional Chinese herbs significantly inhibited the IL-4 induced expression of the low affinity IgE receptor CD23.

In a later study, after 2 months treatment with the Zemaphyte decoction, the number of dendritic cells (RFD1), macrophages (RFD7), T cell (CD25) subsets and Langerhans cells (CD1) were significantly reduced in lesional skin [72]. HLA-DR expression was also significantly decreased and the authors suggest that "clinical improvement of TCHT is associated with significant reduction in antigen presenting cells expressing CD23" [72].

Sheehan and co-workers have clearly demonstrated clinical efficacy of TCHT in adults and children with AD in two double-blind, placebo-controlled trials [68, 73]. In one cross-over trial two groups comprising 40 randomly assigned adults with AD were treated for 2 months. Each group received 4 weeks of the 10-herb mixture and 4 weeks of placebo. In patients who were treated with the active compound there was a significant reduction of erythema and "surface damage", defined by the authors as the net effect of the papulation, vesiculation, scaling, excoriation and lichenification [68]. After the initial trial 17 patients continued treatment for 1 year, of whom 12 had a greater than 90 % and 5 had a greater than 60% reduction in clinical scores [74]. Haematological or biochemical abnormalities did not occur.

Another study by the same group investigated the effectiveness of TCHT in 47 children with severe, non-exudative AD [73]. A standard herbal compound or placebo was given in random order, each for 8 weeks with a 4 weeks washout period in between. The active compound was superior to placebo with a significant decrease in erythema and surface damage scores in the 37 patients who completed the study [73]. Following the initial placebo controlled trial phase, the participants went into an open trial for 1 year with continued treatment. 23 children completed the study of which 18 showed 90% improvement and 5 lesser benefit. Drop-outs were mainly due to lack of response or unpalatability of the preparation [75].

In the short-term study, no haematological, renal or hepatic toxicity was observed, but in the longer term study two children developed elevation of aspartate aminotransferase which reversed within 2 months after stopping treatment. Steroid-like activity of the herbal compound was excluded by analysis of urinary corticosteroid metabolites and great care was taken to establish adequate quality control of the preparation and screening for contamination with other chemicals, which could contribute to serious and unwanted side-effects.

However there are serious concerns about toxicity of TCHT preparations and there are several reports about hepatotoxicity [76–78] including a case report of acute liver failure [77] and also a case of reversible dilated cardiomyopathy [79] following treatment with Chinese herbal compounds. A report of 11 cases of liver damage with an obvious circumstantial link with TCHT suggested that liver damage was most likely to be idiosyncratic rather than dose related.

Immuno-modulators

Interferon therapy

The rationale for using interferon seems to originate from two observations. The first is that the hyper-IgE syndrome appears to respond to interferon α [80]. Secondly, since it appears that IFNγ production is low in atopic subjects, it can be argued that exogenous IFNγ might inhibit production of IgE and activation of Th2 cells.

Interferon α (IFNα)

IFNα has been tried by some investigators. Treatment regimens have been as short as 21 daily injections of 3×10^6 units [81] to thrice weekly for 3 months [82, 83]. Overall, benefits have not been great and have worn off as soon as treatment was stopped. There does not seem to be much place in the future for IFNα in the management of AD.

Interferon γ (IFNγ)

There have been a number of reports of the use of recombinant IFNγ in AD. Reinhold et al. [84] treated 14 patients for 6 weeks. Eight of the 14 (57%) showed marked improvement. In four, the improvement persisted for 3 months after stopping treatment. In a large multi-centre trial, 83 patients were treated in a randomised double-blind placebo-controlled protocol [85]. Daily injections of 50 mg/m^2 were given for 12 weeks. Inevitably, significant placebo effects were obtained on the clinical aspects. However, active treatment was better than placebo. The greatest improvements were seen with erythema (35% improvement *versus* 20% for control), excoriation (35% improvement *versus* 17% for control) and blood eosinophilia. Interestingly, mean IgE levels actually rose by 1100 IU/ml. Although this seems to contradict the rationale the authors argued that it reflects the fact that *in vitro*, higher concentrations of IFNγ are necessary to inhibit IgE production. More recently Stevens et al. have reported the results of up to 2 years treatment with IFNγ [86]. Twenty-four patients self-administered sub-cutaneous IFNγ (50 μg/M^2/day); there was significant improvement in clinical parameters and fall in eosinophil count. Interestingly, again, the IgE levels actually increased. The authors concluded that the treatment is safe but in relation to the relatively modest clinical benefits obtained, this is unlikely to become a widely used form of therapy.

Thymopentin

Thymopentin is the active pentapeptide of an immuno-stimulatory thymic hormone. This substance has been tried in severe AD by a number of groups [87–89]. For example, 16 children were given thrice weekly injections of 50 mg of thymopentin for 6 weeks [88]. After 3 weeks, significant clinical improvement began to take place. The beneficial effect was lost soon after discontinuation of therapy however. There was a decrease in *in vitro* production of IL-4 and an increase in IFNγ production, but serum IgE levels and *in vitro* IgE production were unchanged.

Overall, modification of the biological and immune responses with agents such as interferon or thymopentin is still at the level of a research tool rather than a serious management option.

Allergen avoidance

An attractive and alternative approach in the treatment of atopic dermatitis is to remove the causal factors, which avoids exposing the sufferer to a wide variety of drugs many of which have dangerous and even life-threatening risks associated. One of the big difficulties is the reliable identification of what are the causal or provoking factors in any individual. However, in the geographical areas with temperate climate, as found in Britain, house dust mites and their allergens are a major factor in the provocation of atopic dermatitis.

Anti-house dust mite measures

House dust mites (HDM) are found in all homes in the temperate regions of the earth, and there is evidence that homes of patients with AD have higher levels [90]. The major antigen of HDM (Der p1) is found in their faecal pellets. Many workers have attempted to eradicate HDM and/or other antigens by the use of preparations which kill mites and/or denature their antigens, barrier bedding systems, vacuum cleaning and ventilation (reviewed in [91]).

Methods of killing mites
Manipulation of ambient humidity, liquid nitrogen spray [92], hot washing [93], benzyl benzoate (Acarosan) [91, 94], pirimiphos methyl [95], permethrin [96].

Methods of reducing antigen load or exposure
Mite proof bedding encasements [97, 98], hot washing of bedding [93], benzyl benzoate (Acarosan) [91, 94], NB failure to demonstrate efficacy [97], benzyl tannate complex (Allersearch DMS) [99, 100], NB failure to demonstrate efficacy [98], vacuum cleaning [91, 98].

A number of studies have examined the effect of HDM avoidance in AD. Although they have given the impression that beneficial effects occur, they have mainly been limited in their approach. When patients are admitted to hospital where levels of HDM allergens are low they usually improve [101, 102]. Clark and Adinof [103] showed that in vivo challenge tests suggested environmental allergens as relevant provoking factors in 12 patients. Removal of the patients from the relevant environment was associated with amelioration of symptoms. Most of the studies have instituted partial anti-dust mite measures – either acaricidal sprays with or without high efficiency vacuum cleaners or impervious bags on bedding [100, 104–107]. The problem of using only an acaricide such as liquid nitrogen or benzyl benzoate, is that even though the mites may be killed, the antigens in their bodies and faeces remain.

A number of workers looked at the effects of acaricides on mite antigen load in isolation from the clinical setting [95, 100, 108]. Other workers examined clinical consequences of anti-HDM measures in asthma [109, 110]. Some of these studies did not monitor the effects on HDM antigen (Der p1) levels. Marks et al. [107] showed that the combination of impermeable bed covers and a spray that killed mites and denatured their antigen was not sufficient to cause a sustained reduction in Der p1 level in the home to benefit asthma patients. Ehnert et al. [111] showed that killing mites with benzyl benzoate failed to reduce Der p1 levels although polyurethane bed bags greatly reduced the Der p1 concentration on the bed surface. This was associated with a significant reduction in bronchial hyper-reactivity. Clearly before anti-dust mite measures can be definitely added to the recommended management of AD these correlations between reduction in dust mite antigen levels and clinical improvement must be studied and defined much more thoroughly.

The question of whether anti-HDM measures can benefit AD had received little attention and, until recently, was the subject of conjecture and uncertainty. Roberts [104] and August [105] found improvement in AD after the use of plastic covers and vacuum cleaning and/or removal of carpets. However, the present author and colleagues examined a combination of anti-dust mite measures that would be easily applicable in normal homes, for effects on both Der p1 load and the clinical activity of AD [98]. In a double blind comparison, active treatment comprised microporous Goretex bedding bags (Intervent) applied to the mattress, pillow(s) and top covers, a benzyltannate (acaricide/allergen denaturant, Allersearch DMS) spray for carpets and a high filtration vacuum cleaner.

The main findings are summarised below. Forty-eight patients between 7 and 65 years with AD were assigned at random to receive either active or placebo measures. The trial period was 6 months, during which a specially trained nurse collected baseline dust samples from the beds and carpets in the bedroom and living room, applied the measures and re-sampled the dust at monthly intervals. The active treatment is given above; the placebo measures comprised cotton bed bags, water spray and a low power poor filtration vacuum cleaner with a rigid box construction. The carpets were re-sprayed after 3 months; twice the recommended quantities were applied on each occasion. The patients were assessed at monthly intervals by a blinded observer and the extent and severity of the dermatitis was scored.

The first question to be examined was "How much do the anti-HDM measures reduce antigen levels in bedding and carpets?" The Goretex bed bags reduced Der P1 levels on the surface of the bed by 98% (Fig. 2). The reduction persisted throughout the study period. The effects on Der P1 content of the treated carpets showed two main points (Tab. 2). Firstly, the placebo treatment was just as good as the active treatment – the final Der P1 levels in the carpets of both groups were the same. This indicates that a simple modern vacuum cleaner is just as good as the combination of a high power, high filtration vacuum cleaner plus the acaricidal spray. Secondly, it is clear that there is an irreducible lower limit for mite allergens in carpets, below which even the best cleaning methods cannot go. It is not clear from this study whether this level of Der P1 in carpets is low enough to be clinically irrelevant.

The second question was "What are the clinical consequences of reducing Der P1 in the domestic environment?" There were highly significant clinical benefits – a majority of both children and adults showing improvement (Fig. 3). Some of the worst affected patients showed the biggest clinical improvement.

In conclusion it is now quite clear that measures aimed at eliminating contact with HDM and their antigens, can result in great clinical improvement in many patients with AD. Unfortunately it is difficult to predict which patients will gain the

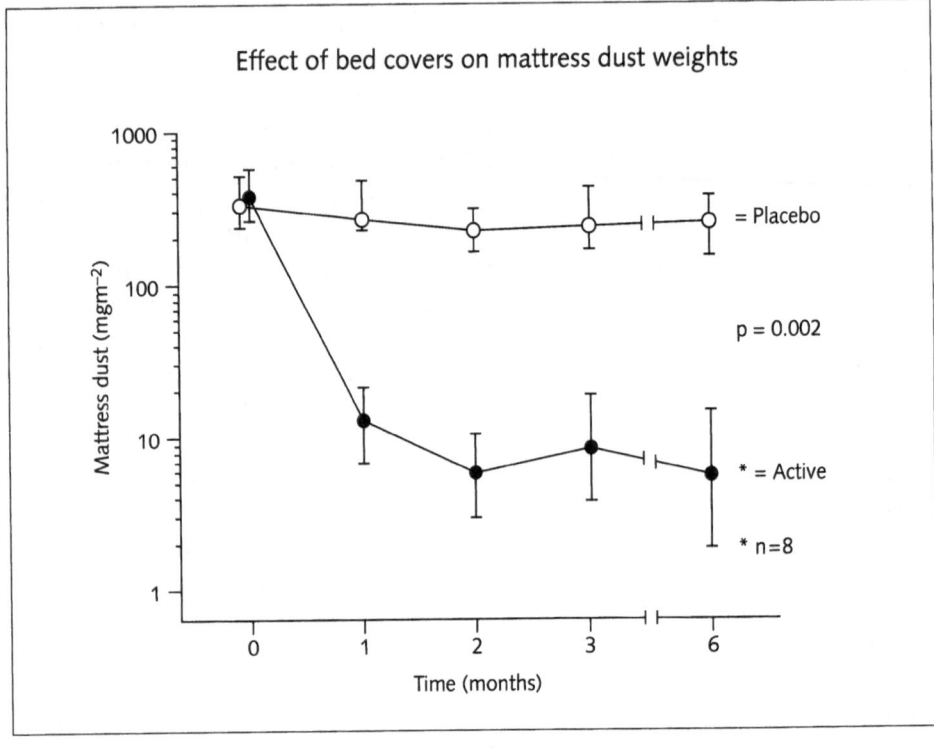

Figure 2
Effect of Goretex bedding encasements on dust load. The active covers resulted in > 95%
reduction in dust weight – too little residual dust to measure Der P1 levels. (Reproduced with
permission from [98].)

Table 2 - The concentration of Der p1 (ng/m^2) in bedroom and living room carpets after use
of either a high power vacuum cleaner plus Allersearch DMC spray or a cheap vacuum clean-
er plus water spray.

Treatment group	Time (months)	bedroom carpet (Der p1ng/m^2)		living room carpet (Der p1ng/m^2)	
Active	0	3669[1]	(16-42014)[2]	1321	(0-30130)
	6	249	(0-15625)	271	(0-4270)
Placebo	0	2161	(32-36756)	913	(5422391)
	6	225	(0-18063)	271	(7-6726)

[1]*Geometric mean values;* [2]*Ranges. (Reproduced with permission from [98] .)*

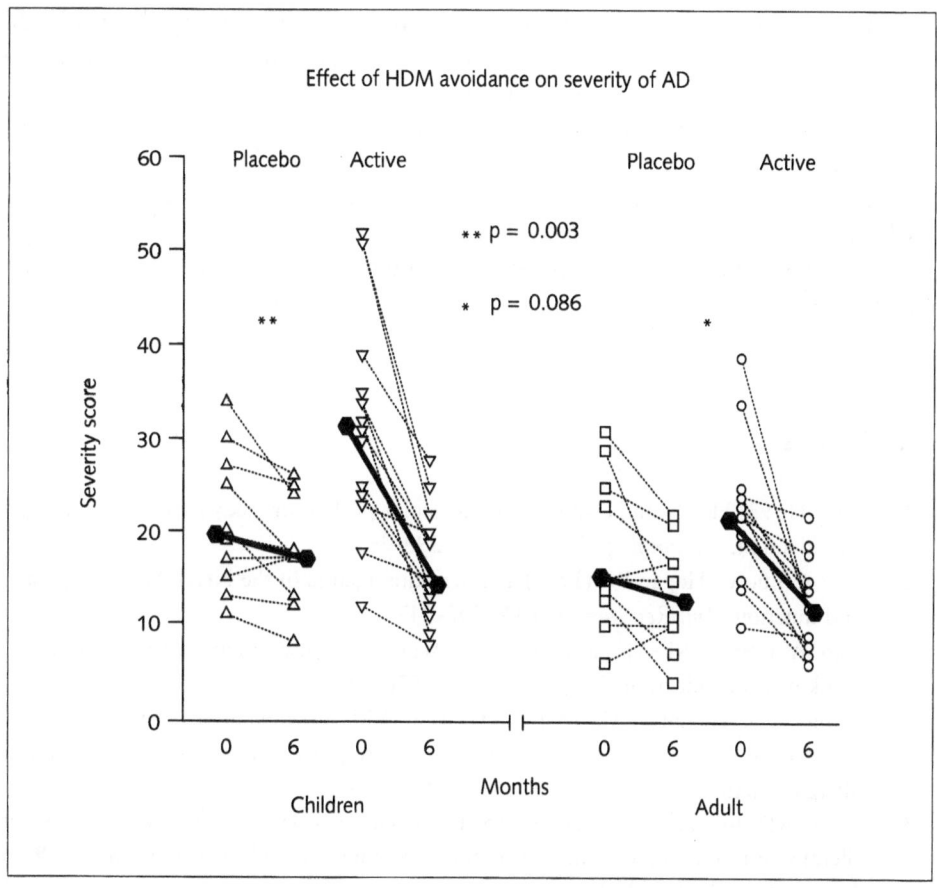

Figure 3
Effect of house dust mite avoidance on severity of atopic dermatitis. Highly significant reductions were seen in eczema severity scores in the groups of both children and adults receiving active measures. Some of the most severe cases showed the biggest improvements. (Reproduced with permission from [112].)

benefit. Even so, since the most severely affected patients showed the best results, it is clearly worth recommending these measures but there are a few caveats. Firstly, it is essential to perform the measures properly. The most important part of the environment for adults with AD is probably the bed, while the floors are also important for children who play on the carpet. The right grade of bed bags must be used – unfortunately the Goretex type used in our study are no longer available in the UK but others are. The mattress must be encased in a bag but it is probably acceptable to put the pillows and top covers through a hot wash and tumble dry every 6–8 weeks. Most modern vacuum cleaners are up to the job of reducing allergen load in carpets. However, removal of carpets and replacement with something like vinyl sheet is the most complete treatment for floors. It is unlikely that the acaricidal products make a significant contribution and measures such as vacuuming the bed is a total waste of time.

There are still many aspects of this problem that require clarification, such as which parts of the house need to be treated. Overall, however, the value of performing dust mite elimination measures for improved control of AD is now clearly established.

References

1   Liu J (1993) FK506 and cyclosporin, molecular probes for studying intracellular signal transduction. *Immunology Today* 14 (6): 290–265

2   Rao A, Luo C, Hogan PG (1997) Transcription factors of the NFAT family: regulation and function. *Ann Rev Immunol* 15: 707–747

3   Rao A (1995) NFATp, a cyclosporine-sensitive transcription factor implicated in cytokine gene induction. *J Leukoc Biol* 57: 536–542

4   Munro CS, Higgins EM, Marks JM, Daly BM, Friedmann PS, Shuster S (1991) Cyclosporin A in atopic dermatitis: therapeutic response is dissociated from effects on allergic reactions. *Br J Dermatol* 124: 43–48

5   Camp RD, Reitamo S, Friedmann PS, Ho V, Heule F (1993) Cyclosporin A in severe, therapy-resistant atopic dermatitis: report of an international workshop, April 1993. *Br J Dermatol* 129: 217–220

6   Munro CS, Levell NJ, Shuster S, Friedmann PS (1994) Maintenance treatment with Cyclosporin in atopic eczema. *Br J Dermatol* 130: 376–80

7   Berth Jones J, Finlay AY, Zaki I, Tan BB, Goodyear H, Lewis-Jones MS, Cork MJ, Bleehen SS, Salek MS, Allen BR et al (1996) Cyclosporine in severe childhood atopic-dermatitis – a multicenter study. *J Am Acad Dermatol* 34: 1016–1021

8   Berth Jones J, Graham-Brown RAC, Marks B, Camp RDR, English JSC, Freeman K, Holden CA, Rogers SCF, Oliwiecki S, Friedmann PS et al (1997) Long-term efficacy and safety of cyclosporin in severe adult atopic dermatitis. *Br J Dermatol* 136: 76–81

9   Bieber T (1998) Topical tacrolimus (FK 506): a new milestone in the management of atopic dermatitis. *J Allergy Clin Immunol* 102: 555–557

10   Nakagawa H, Etoh T, Ishibashi Y, Higaki Y, Kawashima M, Torii H, Harada S (1994) Tacrolimus ointment for atopic dermatitis. *Lancet* 344: 883

11   Aoyama H, Tabata N, Tanaka M, Uesugi Y, Tagami H (1995) Succesful treatment of resistant facial lesions of atopic dermatitis with 0.1% FK506 ointment. *Br J Dermatol* 133: 494–495

12   Alaiti S, Kang SW, Fiedler VC, Ellis CN, Spurlin DV, Fader D, Ulyanov G, Gadgil SD,

Tanase A, Lawrence I et al (1998) Tacrolimus (FK506) ointment for atopic dermatitis: a phase I study in adults and children. *J Am Acad Dermatol* 38: 69–76

13 Boguniewicz M, Fiedler VC, Raimer S, Lawrence ID, Leung DYM, Hanifin JM (1998) A randomized, vehicle-controlled trial of tacrolimus ointment for treatment of atopic dermatitis in children. *J Allergy Clin Immunol* 102: 637–44

14 Ruzicka T, Bieber T, Schopf E, Rubins A, Dobozy A, Bos JD, Jablonska S, Ahmed I, Thestruppedersen K, Daniel F et al (1997) A short-term trial of tacrolimus ointment for atopic dermatitis. *New Engl J Med* 337: 816–821

15 Meingassner JG, Grassberger M, Fahrngruber H, Moore HD, Schuurman H, Stutz A (1997) A novel anti-inflammatory drug, SDZ ASM 981, for the topical and oral treatment of skin diseases: *in vivo* pharmacology. *Br J Dermatol* 137: 568–576

16 Rappersberger K, Meingassner JG, Fialla R, Fodinger D, Sterniczky B, Rauch S, Putz E, Stutz A, Wolff K (1996) Clearing of psoriasis by a novel immunosuppressive macrolide. *J Invest Dermatol* 106: 701–710

17 Hanifin JM, Chan SC, Cheng JB, Tofte SJ, Henderson WR, Kirby DS, Weiner ES (1996) Type 4 phosphodiesterase inhibitors have clinical and *in vitro* anti-inflammatory effects in atopic dermatitis. *J Invest Dermatol* 107: 51–56

18 Meingassner JG, Stutz A (1992) Antiinflammatory effects of macrophilin-interacting drugs in animal-models of irritant and allergic contact-dermatitis. *Int Arch Allergy Immunol* 99: 486–489

19 Ransom JT (1995) Mechanism of action of mycophenolate mofetil. *Therapeutic Drug Monitoring* 17: 681–684

20 Goldblum R (1993) Therapy of rheumatoid-arthritis with mycophenolate mofetil. *Clin Exp Rheumatol* 11: S117–S119

21 Nousari HC, Griffin WA, Anhalt GJ (1998) Successful therapy for bullous pemphigoid with mycophenolate mofetil. *J Am Acad Dermatol* 39: 497–498

22 Haufs MG, Beissert S, Grabbe S, Schutte B, Luger TA (1998) Psoriasis vulgaris treated successfully with mycophenolate mofetil. *Br J Dermatol* 138: 179–181

23 Hanifin JM, Chan SC (1995) Monocyte phosphodiesterase abnormalities and dysregulation of lymphocyte function in atopic-dermatitis. *J Invest Dermatol* 105: S84–S88

24 Chan SC, Kim JW, Henderson WR, Jr., Hanifin JM (1993) Altered prostaglandin E2 regulation of cytokine production in atopic dermatitis. *J Immunol* 151: 3345–3352

25 Betz M, Fox BS (1991) Prostaglandin E2 inhibits production of Th1 lymphokines but not of Th2 lymphokines. *J Immunol* 146: 108–113

26 Safko MJ, Chan SC, Cooper KD, Hanifin JM (1981) Heterologous desensitisation of leukocytes: a possible mechanism for beta adrenergic blockade in atopic dermatitis. *J Allergy Clin Immunol* 68: 218–225

27 Grewe SR, Chan SC, Hanifin JM (1982) Elevated leukocyte cyclic AMP-phosphodiesterase in atopic disease: a possible mechanism for cyclic AMP-agonist hyporesponsiveness. *J Allergy Clin Immunol* 70: 452–457

28 Sawai T, Ikai K, Uehara M (1998) Cyclic adenosine monophosphate phosphodiesterase

activity in peripheral blood mononuclear leucocytes from patients with atopic dermatitis: correlation with respiratory atopy. *Br J Dermatol* 138: 846–848

29 Heskel NS, Chan SC, Thiel ML, Stevens SR, Casperson LS, Hanifin JM (1984) Elevated umbilical-cord blood leukocyte cyclic adenosine-monophosphate phosphodiesterase activity in children with atopic parents. *J Am Acad Dermatol* 11: 422–426

30 Mcmillan JC, Heskel NS, Hanifin JM (1985) Cyclic AMP-phosphodiesterase activity and histamine-release in cord blood leukocyte preparations. *Acta Dermato-Venereologica* 5114: 24–32

31 Holden CA, Chan SC, Hanifin JM (1986) Monocyte localization of elevated cAMP phosphodiesterase activity in atopic dermatitis. *J Invest Dermatol* 87: 372–376

32 Chan SC, Reifsnyder D, Beavo JA, Hanifin JM (1993) Immunochemical characterization of the distinct monocyte cyclic amp-phosphodiesterase from patients with atopic-dermatitis. *J Allergy Clin Immunol* 91: 1179–1188

33 Crocker IC, Ohia SE, Church MK, Townley RG (1998) Phosphodiesterase type 4 inhibitors, but not glucocorticoids, are more potent in suppression of cytokine secretion by mononuclear cells from atopic than nonatopic donors. *J Allergy Clin Immunol* 102: 797–804

34 Gantner F, Tenor H, Gekeler V, Schudt C, Wendel A, Hatzelmann A (1997) Phosphodiesterase profiles of highly purified human peripheral blood leukocyte populations from normal and atopic individuals: a comparative study. *J Allergy Clin Immunol* 100: 527–535

35 Li SH, Chan SC, Kramer SM, Hanifin JM (1993) Modulation of leukocyte cyclic AMP phosphodiesterase activity by recombinant interferon-gamma: evidence for a differential effect on atopic monocytes. *J Interferon Res* 13: 197–202

36 Spina D, Landells LJ, Page CP (1998) The role of theophylline and phosphodiesterase 4 isoenzyme inhibitors as anti-inflammatory drugs. *Clin Exp Allergy* 28: 24–34

37 Cooper KD, Kang K, Chan SC, Hanifin JM (1985) Phosphodiesterase inhibition by Ro20-1724 reduces hyper-IgE synthesis by atopic dermatitis cells *in vitro*. *J Invest Dermatol* 84: 477–482

38 Chan SC, Li SH, Hanifin JM (1993) Increased interleukin-4 production by atopic mononuclear leukocytes correlates with increased cyclic adenosine monophosphate-phosphodiesterase activity and is reversible by phosphodiesterase inhibition. *J Invest Dermatol* 100: 681–684

39 Tenor H, Hatzelmann A, Church MK, Schudt C, Shute JK (1996) Effects of theophylline and rolipram on leukotriene C-4 (LTC(4) synthesis and chemotaxis of human eosinophils from normal and atopic subjects. *Brit J Pharmacol* 118: 1727–1735

40 Lidington E, Nohammer C, Dominguez M, Ferry B, Rose ML (1996) Inhibition of the transendothelial migration of human lymphocytes but not monocytes by phosphodiesterase inhibitors. *Clin Exp Immunol* 104: 66–71

41 Essayan DM, Huang SK, Kageysobotka A, Lichtenstein LM (1995) Effects of nonselective and isozyme-selective cyclic-nucleotide phosphodiesterase inhibitors on antigen-

induced cytokine gene-expression in peripheral-blood mononuclear-cells. *Am J Resp Cell Mol Biol* 13: 692–702

42 Banner KH, Page CP (1995) Theophylline and selective phosphodiesterase inhibitors as antiinflammatory drugs in the treatment of bronchial-asthma. *Eur Resp J* 8: 996–1000

43 Massey WA (1993) Pathogenesis and pharmacological modulation of the cutaneous late- phase reaction. *Ann Allergy* 71: 578–584

44 Talbot SF, Atkins PC, Goetzl EJ, Zweiman B (1985) Accumulation of leukotriene C4 and histamine in human allergic skin reactions. *J Clin Invest* 76: 650–656

45 Fauler J, Neumann C, Tsikas D, Frolich JC (1993) Enhanced synthesis of cysteinyl leukotrienes in atopic-dermatitis. *Br J Dermatol* 128: 627–630

46 Sansom JE, Taylor GW, Dollery CT, Archer CB (1997) Urinary leukotriene E-4 levels in patients with atopic dermatitis. *Br J Dermatol* 136: 790–791

47 Carucci JA, Washenik K, Weinstein A, Shupack J, Cohen DE (1998) The leukotriene antagonist zafirlukast as a therapeutic agent for atopic dermatitis. *Arch Dermatol* 134: 785–786

48 Parrish JA, Jaenicke KF (1981) Action spectrum for phototherapy of psoriasis. *J Invest Dermatol* 76(5): 359–362

49 Jekler J, Larko O (1988) UVB phototherapy of atopic dermatitis. *Br J Dermatol* 119 (6): 697–705

50 Jekler J, Larko O (1990) Combined UVA-UVB versus UVB phototherapy for atopic dermatitis: A paired comparison study. *J Am Acad Dermatol* 22: 49–53

51 Wulf HC, Bech Thomsen N (1998) A UVB phototherapy protocol with very low dose increments as a treatment of atopic dermatitis. *Photodermatol Photoimmunol Photomed* 14 (1): 1–6

52 Midelfart K, Stenvold SE, Volden G (1985) Combined UVB and UVA phototherapy of atopic eczema. *Dermatologica* 171 (2): 95–98

53 George SA, Bilsland DJ, Johnson BE, Ferguson J (1993) Narrow-band (TL-01) UVB air-conditioned phototherapy for chronic severe adult atopic dermatitis. *Br J Dermatol* 128: 49–56

54 Collins P, Ferguson J (1995) Narrowband (TL-01) UVB air-conditioned phototherapy for atopic eczema in children. *Br J Dermatol* 133 (4): 653–655

55 Krutmann J, Czech W, Diepgen T, Niedner R, Kapp A, Schopf E (1992) High-dose UVA1 therapy in the treatment of patients with atopic dermatitis. *J Am Acad Dermatol* 26: 225–230

56 Grabbe J, Welker P, Humke S, Grewe M, Schopf E, Henz BM, Krutmann J (1996) High-dose ultraviolet A1 (UVA1), but not UVA/UVB therapy, decreases IgE-binding cells in lesional skin of patients with atopic eczema. *J Invest Dermatol* 107 (3): 419–422

57 Krutmann J, Diepgen TL, Luger TA, Grabbe S, Meffert H, Sonnichsen N, Czech W, Kapp A, Stege H, Grewe M et al (1998) High-dose UVA1 therapy for atopic dermatitis: results of a multicenter trial. *J Am Acad Dermatol* 38 (4): 589–593

58 Morison WL, Parrish J, Fitzpatrick TB (1978) Oralen psoralen photochemotherapy of atopic eczema. *Br J Dermatol* 98 (1): 25–30

59   Yoshiike T, Aikawa Y, Sindhvananda J, Ogawa H (1993) A proposed guideline for psoralen photochemotherapy (PUVA) with atopic dermatitis: successful therapeutic effect on severe and intractable cases. *J Dermatol Sci* 5: 50–53

60   Atherton DJ, Carabott F, Glover MT, Hawk JL (1988) The role of psoralen photochemotherapy (PUVA) in the treatment of severe atopic eczema in adolescents. *Br J Dermatol* 118 (6): 791–795

61   Sheehan MP, Atherton DJ, Norris P, Hawk J (1993) Oral psoralen photochemotherapy in severe childhood atopic eczema: an update. *Br J Dermatol* 129: 431–436

62   Lindelof B, Sigurgeirsson B, Tegner E, Larko O, Johannesson A, Berne B, Christensen OB, Andersson T, Torngren M, Molin L et al (1991) PUVA and cancer: a large-scale epidemiological study. *Lancet* 338 (8759): 91–93

63   Kaidbey KH, Kligman AM (1981) Cumulative effects from repeated exposures to ultraviolet radiation. *J Invest Dermatol* 76 (5): 352–355

64   Parrish JA, Zaynoun S, Anderson RR (1981) Cumulative effects of repeated subthreshold doses of ultraviolet radiation. *J Invest Dermatol* 76 (5): 356–358

65   Storbeck K, Holzle E, Schurer N, Lehmann P, Plewig G (1993) Narrow-band UVB (311 nm) versus conventional broad-band UVB with and without dithranol in phototherapy for psoriasis. *J Am Acad Dermatol* 28 (2 Pt 1): 227–231

66   van Iperen HP, Beijersbergen van Henegouwen GMJ (1997) Clinical and mechanistic aspects of photopheresis. *J Photochem Photobiol B-Biology* 93: 99–109

67   Richter HI, Billmann Eberwein C, Grewe M, Stege H, Berneburg M, Ruzicka T, Krutmann J (1998) Successful monotherapy of severe and intractable atopic dermatitis by photopheresis. *J Am Acad Dermatol* 38(4): 585–588

68   Sheehan MP, Rustin MH, Atherton DJ, Buckley C, Harris DW, Brostoff J, Ostlere L, Dawson A, Harris DJH (1992) Efficacy of traditional Chinese herbal therapy in adult atopic dermatitis. *Lancet* 340: 13–17

69   Latchman Y, Bannerjee P, Poulter LW, Rustin MH, Brostoff J (1996) Association of immunological changes with clinical efficacy in atopic eczema patients treated with traditional Chinese herbal therapy (Zemaphyte). *Int Arch Allergy Immunol* 109: 243–249

70   Latchman Y, Whittle B, Rustin MH, Atherton DJ, Brostoff J (1994) The efficacy of traditional Chinese herbal therapy in atopic eczema. *Int Arch Allergy Immunol* 104 (3): 222–226

71   Latchman Y, Bungy GA, Atherton DJ, Rustin MH, Brostoff J (1995) Efficacy of traditional Chinese herbal therapy *in vitro*. A model system for atopic eczema: inhibition of CD23 expression on blood monocytes. *Br J Dermatol* 132 (4): 592–598

72   Xu XJ, Banerjee P, Rustin MH, Poulter LW (1997) Modulation by Chinese herbal therapy of immune mechanisms in the skin of patients with atopic eczema. *Br J Dermatol* 136 (1): 54–59

73   Sheehan MP, Atherton DJ (1992) A controlled trial of traditional Chinese medicinal plants in widespread non-exudative atopic eczema. *Br J Dermatol* 126: 179–184

74   Sheehan MP, Stevens H, Ostlere LS, Atherton DJ, Brostoff J, Rustin MH (1995) Follow-

up of adult patients with atopic eczema treated with Chinese herbal therapy for 1 year. *Clin Exp Dermatol* 20 (2): 136–140

75 Sheehan MP, Atherton DJ (1994) One-year follow up of children treated with Chinese medicinal herbs for atopic eczema. *Br J Dermatol* 130: 488–493

76 Graham-Brown RAC (1992) Toxicity of Chinese herbal remedies. *Lancet* 340: 673

77 Perharic-Walton L, Murray V (1992) Liver failure after traditional Chinese herbal therapy. *Lancet* 340: 674

78 Perharic L, Shaw D, Leon C, De Smet PA, Murray VS (1995) Possible association of liver damage with the use of Chinese herbal medicine for skin disease. *Vet Hum Toxicol* 37: 562–566

79 Ferguson JE, Chalmers RJ, Rowlands DJ (1997) Reversible dilated cardiomyopathy following treatment of atopic eczema with Chinese herbal medicine. *Br J Dermatol* 136 (4): 592–593

80 Souillet G, Rousset F, de Vries JE (1989) Alpha-interferon treatment of patient with hyper IgE syndrome [letter]. *Lancet* 1: 1384

81 Nielsen BW, Reimert CM, Hammer R, Schiotz PO, Thestrup Pedersen K (1994) Interferon therapy for atopic dermatitis reduces basophil histamine release, but does not reduce serum IgE or eosinophilic proteins. *Allergy* 49: 120–128

82 Gruschwitz MS, Peters KP, Heese A, Stosiek N, Koch HU, Hornstein OP (1993) Effects of interferon-alpha-2b on the clinical course, inflammatory skin infiltrates and peripheral blood lymphocytes in patients with severe atopic eczema. *Int Arch Allergy Immunol* 101: 20–30

83 Torrelo A, Harto A, Sendagorta E, Czarnetzki BM, Ledo A (1992) Interferon-alpha therapy in atopic dermatitis. *Acta Derm Venereol (Stockh)* 72: 370–372

84 Reinhold U, Kukel S, Brzoska J, Kreysel HW (1993) Systemic interferon gamma treatment in severe atopic dermatitis. *J Am Acad Dermatol* 29: 58–63

85 Hanifin JM, Schneider LC, Leung DYM, Ellis CN, Jaffe HS, Izu AE, Bucalo LR, Hirabayashi SE, Tofte SJ, Cantugonzales G et al (1993) Recombinant interferon gamma-therapy for atopic-dermatitis. *J Am Acad Dermatol* 28: 189–197

86 Stevens SR, Hanifin JM, Hamilton T, Tofte SJ, Cooper KD (1998) Long-term effectiveness and safety of recombinant human interferon gamma therapy for atopic dermatitis despite unchanged serum IgE levels. *Arch Dermatol* 134: 799–804

87 Harper JI, Mason UA, White TR, Staughton RC, Hobbs JR (1991) A double-blind placebo-controlled study of thymostimulin (TP-1) for the treatment of atopic eczema. *Br J Dermatol* 125: 368–372

88 Hsieh KH, Shaio MF, Liao TN (1992) Thymopentin treatment in severe atopic dermatitis--clinical and immunological evaluations. *Arch Dis Child* 67: 1095–1102

89 Stiller MJ, Shupack JL, Kenny C, Jondreau L, Cohen DE, Soter NA (1994) A double-blind, placebo-controlled clinical-trial to evaluate the safety and efficacy of thymopentin as an adjunctive treatment in atopic-dermatitis. *J Am Acad Dermatol* 30: 597–602

90 Colloff MJ (1992) Exposure to house dust mite in homes of people with atopic dermatitis. *Br J Dermatol* 127: 322–327

91 Colloff MJ, Ayres J, Carswell F, Howarth PH, Merrett TG, Mitchell EB, Walshaw MJ, Warner JO, Warner JA, Woodcock AA (1992) The control of allergens of dust mites and domestic pets: a position paper. *Clin Exp Allergy* 22 (Supp 2): 1–28

92 Dorward AJ, Colloff MJ, Mackay NS, McSharry C, Thompson NC (1988) Effect of house dust mite avoidance measures in adult atopic asthma. *Thorax* 43: 98–102

93 Andersen A, Rosen J (1989) House dust mite Dermatophagoides pteronyssinus and its allergens: effects of washing. *Allergy* 44: 396–400

94 Morrow Brown H, Merrett TG (1991) Effectiveness of an acaricide in management of house dust mite allergy. *Ann Allergy* 67: 25–31

95 Mitchell EB, Wilkins S, McCallum Deighton J, Platts-Mills TAE (1985) Reduction of house dust mite allergen levels in the home: use of the acaricide, pirimiphos methyl. *Clin Allergy* 15: 235–240

96 Cameron MM (1997) Can house dust mite-triggered atopic dermatitis be alleviated using acaricides? *Br J Dermatol* 137: 1–8

97 Weeks J, Oliver J, Birmingham K, Crewes A, Carswell F (1995) A combined approach to reduce mite allergen in the bedroom. *Clin Exp Allergy* 25: 1179–1183

98 Tan BB, Weald D, Strickland I, Friedmann PS (1996) Double-blind controlled trial of effect of housedust-mite allergen avoidance on atopic dermatitis. *Lancet* 347: 15–18

99 Green WF, Nicholas NR, Salome CM, Woolcock AJ (1989) Reduction of house dust mite and mite allergens: effects of spraying carpets and blankets with Allersearch DMS, an acaricide combined with an allergen reducing agent. *Clin Exp Dermatol* 19: 203–207

100 Warner JA, Marchant JL, Warner JO (1993) Allergen avoidance in the homes of asthmatic children: the effect of Allersearch DMS. *Clin Exp Allergy* 23: 279–286

101 Platts-Mills TAE, Mitchell EB, Rowntree S, Chapman MD, Wilkins SR (1983) The role of house dust mite allergens in atopic dermatitis. *Clin Exp Dermatol* 8: 233–247

102 Sanda T, Yasue T, Oohashi M, Yasue A (1992) Effectiveness of house dust-mite allergen avoidance through clean room therapy in patients with atopic dermatitis. *J Allergy Clin Immunol* 89: 653–657

103 Clark RA, Adinoff AD (1989) Aeroallergen contact can exacerbate atopic dermatitis: patch tests as a diagnostic tool. *J Am Acad Dermatol* 21: 863–869

104 Roberts DLL (1984) House dust mite avoidance and atopic dermatitis. *Br J Dermatol* 110: 735–736

105 August PJ (1984) House dust mite causes atopic eczema. A preliminary study. *Br J Dermatol* 111 (26): 10–11

106 Denman AM, Cornthwaite D (1990) Control of house dust mite in bedding. *Lancet* 335: 1038

107 Marks GB, Tovey ER, Green W, Shearer M, Salome CM, Woolcock AJ (1994) House dust mite allergen avoidance: a randomised controlled trial of surface chemical treatment and encasement of bedding. *Clin Exp Allergy* 24: 1078–1083

108 Kalra S, Crank P, Hepworth J, Pickering CAC, Woodcock AA (1993) Concentrations of domestic house dust mite allergen Der p1 after treatment with solidified benzyl benzoate (Acarosan) or liquid nitrogen. *Thorax* 48: 10–13

109 Murray AB, Ferguson AC (1983) Dust free bedrooms in the treatment of asthmatic children with house dust mite allergy: a controlled trial. *Pediatrics* 71: 418–422

110 Walshaw MJ, Evans CC (1986) Allergen avoidance in house dust mite sensitive adult asthma. *QJM* 58: 199–215

111 Ehnert B, Lau-Schadendorf S, Weber A, Buettner P, Schou C, Wahn U (1992) Reducing domestic exposure to house dust mite allergen reduces bronchial hyperreactivity in sensitive children with asthma. *J Allergy Clin Immunol* 90: 135–138

112 Friedmann PS, Tan BB, Musaba E, Strickland I (1995) Pathogenesis and management of atopic dermatitis. *Clin Exp Allergy* 25: 799–806

Index